The MACAT Library
世界思想宝库钥匙丛书

解析蕾切尔·卡森
《寂静的春天》

AN ANALYSIS OF
RACHEL CARSON'S
SILENT SPRING

Nikki Springer ◎ 著
祝平 ◎ 译

上海外语教育出版社
SHANGHAI FOREIGN LANGUAGE EDUCATION PRESS

目　录

引 言 1
    蕾切尔·卡森其人 2
    《寂静的春天》的主要内容 3
    《寂静的春天》的学术价值 4

第一部分：学术渊源 7
    1. 作者生平与历史背景 8
    2. 学术背景 11
    3. 主导命题 16
    4. 作者贡献 20

第二部分：学术思想 23
    5. 思想主脉 24
    6. 思想支脉 29
    7. 历史成就 33
    8. 著作地位 37

第三部分：学术影响 41
    9. 最初反响 42
    10. 后续争议 47
    11. 当代印迹 52
    12. 未来展望 56

术语表 60
人名表 66

CONTENTS

WAYS IN TO THE TEXT	71
Who Was Rachel Carson?	72
What Does *Silent Spring* Say?	73
Why Does *Silent Spring* Matter?	75
SECTION 1: INFLUENCES	77
Module 1: The Author and the Historical Context	78
Module 2: Academic Context	82
Module 3: The Problem	88
Module 4: The Author's Contribution	94
SECTION 2: IDEAS	99
Module 5: Main Ideas	100
Module 6: Secondary Ideas	106
Module 7: Achievement	111
Module 8: Place in the Author's Work	115
SECTION 3: IMPACT	121
Module 9: The First Responses	122
Module 10: The Evolving Debate	127
Module 11: Impact and Influence Today	133
Module 12: Where Next?	138
Glossary of Terms	143
People Mentioned in the Text	150
Works Cited	153

引 言

要点

- 蕾切尔·卡森是 20 世纪最具影响力的环境*作家之一。
- 其著作《寂静的春天》被广泛誉为"点燃美国环境运动*的星星之火"。
- 《寂静的春天》已经出版 50 多年了,至今依然是环境专业学生和专业人员的重要读本。

蕾切尔·卡森其人

蕾切尔·卡森(1907—1964)是美国环境作家,人们通常认为其极具影响力的著作《寂静的春天》(1962)引发了美国 20 世纪六七十年代的环境运动(一场社会政治运动,强调保护自然界的重要性,呼吁认识自然资源的有限性)。

卡森的母亲对她影响很大,教育、鼓励她建立与自然界的深层联系,尊重并热爱自然界。卡森成长于宾夕法尼亚州的一个家庭农场,后考入宾夕法尼亚女子学院并获得奖学金。她最初主修英语,不过很快就转了专业,1929 年毕业时获得了生物学*(研究生命有机体的科学)学士学位。接着,她又到伍兹霍尔研究所和约翰·霍普金斯大学深造,并于 1932 年获得约翰·霍普金斯大学动物学*(研究动物生命的科学)硕士学位。[1] 卡森作为科学家和撰稿人在美国政府工作了 15 年,此后成为独立的专职作家。

卡森的作品,尤其是《寂静的春天》,不仅在发动环境运动方面起到了核心作用,而且对公众意识产生了独特影响,使公众意识到对杀虫剂*进行严格监管的必要性。杀虫剂是为了控制对作物健康造成威胁的害虫（尤其是昆虫*）而使用的化学品。《寂静的春天》因而受到了农业及杀虫剂产业的猛烈攻击。作为回应，卡森指出了政治漏洞是如何容许这些产业不顾其产品及生产方式所导致的环境后果而进行生产的。但此类回应未能持续很长时间，因为卡森1964年因乳腺癌辞世，年仅56岁。

《寂静的春天》的主要内容

对于任何一位希望开阔思想和见识的地球公民而言，《寂静的春天》都不啻为一个关乎世界及其生态系统*（各种植物、动物及其栖息地组成的系统）的警世寓言。数百万学者、政策制定者及有关人士已在它的帮助下认识到人类对自然的影响，并开始思考如何找到可持续的生活方式来保护当代人及后世子孙。

《寂静的春天》一书旨在讲述人类活动对环境造成的负面、广泛而持久的影响，卡森用农用化学杀虫剂这一重要的案例研究证明了这一点。她详细描述了第一次世界大战*（1939—1945）后美国农民广泛使用合成化学杀虫剂所造成的有害的、毁灭性的后果——从人类癌症*和出生缺陷到小动物（尤其是鸟类）的死亡，无所不包。事实上，书名中的"寂静"指的就是杀虫剂毁灭了野生动植物后再无鸟鸣的情形。卡森呼吁将这类化学品更名为"杀生物剂"*——她相信这是对其威力和危险更为准确的描述。人们通常认为，1972年美国禁用滴滴涕*（DDT，化学名为"二氯二苯三氯乙烷"的杀虫剂），应归功于《寂静的春天》所发出的警示。

美国政府一方面补贴农民生产出过剩的农产品，同时又宣称必须使用化学杀虫剂才能确保充足的食品供应，卡森认为这颇具讽刺意味。她还批评了当时的政治制度，因为这种制度造成了这样一种现状：重视工业巨头的要求，轻视独立科学家和医生所做的关于化学品安全性及其对人类和野生动植物健康影响的研究。

《寂静的春天》最终倡导的是生物性虫害控制*。卡森建议利用生态系统的自然制衡来控制虫害——比如，实行作物种类多样化种植（扩大作物种植品种）而不是单品种种植*（大面积种植单一作物）。她相信这样可以确保食物链的安全，同时避免合成杀虫剂*的危害。卡森关于以更加环境友好的方式控制农业害虫的理念至关重要，直接否定了化学工业声称的"若想要有充足的、可获利的食物供应，污染就不可避免"的说法。她还呼吁采用完全不同的方式使用杀虫剂，即从她所认为的"无差别"使用转为谨慎地限制使用量和使用范围。

卡森关于化学杀虫剂毒害影响的研究背后是一个更大的命题：我们是否应该为了自己的短浅目标去控制环境及其生态系统，而置这一行为对野生动植物、大自然、地球健康或后代的安全与福祉所造成的长期影响于不顾？

《寂静的春天》的学术价值

《寂静的春天》通常被誉为最重要的现代环境文学作品，其中心作用在于提醒公众警惕杀虫剂的使用所造成的危险以及向环境中释放长期效果尚未完全明确的化学品的风险。在科学领域内，它同样具有非常重要的影响。[2]

《寂静的春天》的出版对美国的环境治理产生了深远影响。除了促使美国农业最终禁用滴滴涕外，1970年美国环境保护署（EPA）*的设立也是明显受了卡森著作的影响。环境保护署的设立有助于将对杀虫剂等化学品的监管与农业补贴（包括常被称作"农业巨无霸"的大型企业内部的补贴)分离开来。在此之前,这两项职能均由美国农业部承担,但卡森认为这实际上存在利益冲突。

卡森及其作品持续影响着当今的政策和政治家们。1994年,时任美国副总统艾尔·戈尔*在为《寂静的春天》所作的序言中说,挂在他办公室里的卡森画像不断地提醒自己记住她的著作及其重要意义。[3]

卡森及其作品的影响远远超出了杀虫剂的使用问题。她质疑了大公司与政府监管机构之间的关系是否值得公众信任,那些未被独立研究证实却常被当作事实展现给大众的、以偏概全的说辞是否真实。对于生活在战后相对舒适社会中的人而言,《寂静的春天》所揭露的现实令人震惊。他们在当地五金店买到的化肥包装袋上印着孩子们在修剪完美的草坪上快乐嬉戏的景象,而书中呈现的危险与这样的营销形象相去甚远。如果卡森的断言为真,那么我们身边还会不会有其他这类恐怖的场景存在呢?

卡森同样也为其他女性科学家铺平了道路,成为生态女性主义*运动的先驱。生态女性主义是在《寂静的春天》出版后不久开始的一场政治思潮,主张将环境保护*问题与争取性别平等的原则相互结合。在今天,卡森依然是环境保护者的榜样,她使人们时刻铭记:仅凭一己之力勇敢发声,也能对抗哪怕最强大的利益集团。

---

1. 琳达·李尔:《蕾切尔·卡森——自然的见证者》,纽约:麦克米伦出版社,1998年,第72页。
2. 约书亚·罗斯曼:"蕾切尔·卡森的自然历史",《纽约客》,2012年9月27日。
3. 艾尔·戈尔:《寂静的春天》序言,蕾切尔·卡森著,纽约:霍顿·米夫林出版社,1994年,第XVIII页。

第一部分：学术渊源

1 作者生平与历史背景

要点

- 在家庭农场的成长经历影响了卡森,使她得以深刻地了解自然界。
- 卡森的英语专业背景及其对生物学专业的深入学习,使她能够清晰地向广大读者解释复杂的科学思想。
- 《寂静的春天》于1962年在美国出版,是政治家和公众最早接受的环境文学作品之一。

为何要读这部著作?

蕾切尔·卡森是20世纪环境文学领域最著名、被引用次数最多的作家之一。其著作《寂静的春天》(1962)是所有生态学*(生物学的分支,关注生物群落与环境的关系)专业学生的必读书目——无论他们是否认同卡森的结论。

虽然书中所展示的大部分科学数据和研究都集中于农业杀虫剂的影响,但其主题和经验教训的适用面十分广泛。事实上,《寂静的春天》在今天仍然与50多年前出版时一样与现实密切相关。所以,有关卡森及《寂静的春天》的著作和学术论文数量一直持续上升。

同样,《寂静的春天》今天依然是以一己之力、凭一种声音对抗世界性问题的最佳典范。卡森虽然在《寂静的春天》出版后不久就辞世了,但她仍继续在向我们昭示一个普通人何以能够改进我们所处的社会和星球。她是美国及世界各国现代环境文学的先驱,也是现代环境运动的催化剂。

> "如果我们希望给子孙后代留下一个像样的世界,我们万不可自私或胆怯。"
>
> ——美国前总统吉米·卡特,1977 年 4 月 18 日的电视讲话

作者生平

蕾切尔·卡森(1907—1964)毕生都保持着与自然世界的联系,这种联系是她童年时在美国宾夕法尼亚州斯普林代尔区的 65 英亩自家农场里建立的。她是个书迷,有娴熟的写作技巧,11 岁时就在儿童杂志《圣尼古拉斯》上发表了第一个短篇作品。她大部分的阅读和写作都以自然为主题。她对海洋尤感兴趣,后来"海洋"也成了她职业生涯中的一个写作主题。[1]

卡森学业优异,1925 年中学毕业时成绩名列前茅。此后,她进入位于匹兹堡的宾夕法尼亚女子学院(现查塔姆大学)学习。她一开始就读于英语专业,不久转到了生物学专业,但她仍继续为学校的报纸和文学杂志撰稿。卡森 1929 年以优等成绩大学毕业,接着到位于马里兰州巴尔的摩市的约翰·霍普金斯大学深造,并于 1932 年获得动物学硕士学位。[2]

她本已开始在约翰·霍普金斯大学攻读博士学位,但由于经济原因,被迫于 1934 年辍学并开始工作。不久,卡森的父亲去世,她承担起了赡养母亲的全部责任。在一位大学导师的鼓励下,她向美国渔业局*(现美国鱼类及野生动植物管理局,即美国内政部主要负责野生生物保护的机构)申请了一个临时职位,主要任务是为广播节目撰稿。她也开始定期为《巴尔的摩太阳报》博物学专栏撰稿。她于 1937 年 7 月发表在美国重要杂志《大西洋月刊》上的文章《海底》是她首部引起关注的作品,文章以优美的笔触描写了大

洋洋底的奇妙世界。该文给读者留下了极为深刻的印象，卡森因此受到了出版社邀请，将其扩展成一本书。[3]

卡森的职业生涯十分成功，1949年就成为美国渔业局出版物的主编，但同时仍继续进行独立创作和出版。她的前三部长篇著作《海风下》（1941）、《我们周围的海洋》（1951）和《海的边缘》（1955）都展现了她对自然界（尤其是对海洋及水生生物）的研究兴趣。

创作背景

对卡森一生最重要的早期影响均植根于她在自家农场的成长经历，这段经历让她对自然界产生了发自内心的热爱和尊重。她明白自然界是以神秘复杂的形式在运转，因而重要的是，要尽量按照自然界本身的方式去理解自然，而不是单纯为人类所需而试图去主导它、控制它。

原本只是因失去亲人被迫终止学业、承担养家糊口责任而申请的美国渔业局临时职位，最后竟开启了卡森长达15年的政府工作生涯。在她有关环境主题的著作《我们周围的海洋》和《海的边缘》大获成功后，她终于具备了足够的经济能力以辞去政府工作，从而将时间和精力全部投入到写作中去。

---

1. 琳达·李尔：《蕾切尔·卡森——自然的见证者》，纽约：麦克米伦出版社，1998年，第120页。
2. 李尔：《蕾切尔·卡森》，第63页。
3. 李尔：《蕾切尔·卡森》，第88页。

2 学术背景

要 点

- 在《寂静的春天》中,卡森描述了美国广泛使用化学杀虫剂的危险性。
- 在环境文学草创时期,《寂静的春天》与作家及环保人士奥尔多·利奥波德*的《沙乡年鉴》一道被公认为20世纪环境文学的奠基之作。
- 卡森创作该书,得益于她的乡村成长经历和科学写作训练。

著作语境

卡森的《寂静的春天》出版于二战后科学获得重大进展之时。那时的政府和工业企业都全心接纳进步理念并形成了一种崇尚技术、科学及现代化的文化。清洁、标准化、技术进步、质量控制、工业化大规模生产等均被视为社会进步的标志。从人类登上月球到加工好的盒式快餐的流行,方方面面都反映了人类的进步和对自然界的控制这两大相互交织的主题。

工程领域追求科学精确性和高度技术化的文化价值观也影响着社会对自然界的态度,于是就出现了维护精良、审美别致的独特景观(即依据某些审美观念培育和塑造的景观)、消灭害虫、农业生产效率最大化这些现象。"科技以及在科技领域工作的人们被尊为自由世界的救星和繁荣的保证,"卡森的传记作者琳达·李尔说道。[1]

导致这种心态的原因之一是二战期间开发的军用技术和化学品的民用化,其中包括化学杀虫剂以及与虫媒传染病防治相关的医疗进步。另一个因素是美国巨型商业公司的崛起和工业对政府监管的掌控（即工业企业对政府设定的行业限令的操控）。企业被视为集聚智慧、值得信赖、科技发达的实体，是20世纪50年代美国战后经济发展之源。

> "就算没有激发一代（环境）活动家，卡森也会作为美国文人中最伟大的自然作家之一引领风骚。"
> ——彼得·马塞森："最有影响的100位世纪人物"，《时代周刊》

学科概览

现代环境运动起因于工业革命*引发的污染急剧增加。18世纪后期，工业革命发端于英国，带来技术和工业的高速发展，西方社会也由此从农业经济转型为工业经济。技术进步一方面使经济获得前所未有的增长，为劳动阶层提供了大量新的就业机会。另一方面，它所造成的环境和健康问题也显而易见，主要表现为空气、水和总体环境质量的恶化，人类的健康受到了损害。这种污染的增长一直持续到20世纪。

美国自然资源保护运动*引发了人们对自然界及其保存与维护的政治、文化关注。该运动建立在作家和政治活动家亨利·大卫·梭罗*、诗人拉尔夫·沃尔多·爱默生*、苏格兰环境哲学家约翰·缪尔*等人的著作基础之上，大约从1890年持续至1920年。[2] 它也因1901—1909年在任的美国总统西奥多·"泰迪"·罗斯福*得以普及。罗斯福支持创设美国林务局（一个涉及国家森林管理的机构），并签署了文物保护法案，该法案赋予总统创设国家纪念园区（具有独特的文化或自然价值的地方）的权力。他创设了五个国家公园，这些地区因其独特的属性而被保护，免遭开发利用。[3] 然而，1914—1918 年和 1939—1945 年的两次世界大战以及 20 世纪 20 年代至 30 年代的大萧条带来的灾难性经济下滑使人们将重心从环境转移到了其他方面。直到 20 世纪 50 年代末和 60 年代，环境运动领袖们才又开始关注经济增长、郊区开发、家长式公司（为国民指定解决方案和行为方式的企业机构）、现代化与机械化这些因素对自然界的影响。可以肯定的是，这种关注部分是由于卡森的著作。

在 20 世纪 50 年代卡森创作《寂静的春天》之时，环境文学作品还极为鲜见。但有一个重要的例外，那就是作家与环保人士奥尔多·利奥波德 1949 年出版的《沙乡年鉴——零散素描》。在书中，利奥波德创造了"土地伦理"*这个名词并阐述了其内涵。"土地伦理"是一种主张人类有责任尊重自己与自然及土地景观关系的哲学。《寂静的春天》后来激发了理查德·尼克松总统执政期间（1969—1974）多项联邦环境法案的实施，例如《洁净水法案》（1972）*、《洁净空气法案》（1970）*和《国家环境保护法案》（1970）*。这些法案保护了美国人的健康和环境。

学术渊源

《寂静的春天》是从卡森有关化学杀虫剂的其他研究成果发展而来的。二战以后，美国军方为将化学杀虫剂用于农业而资助了一批研发项目。其中一个项目是在中大西洋地区（美国东海岸纽约附近）消灭吉普赛飞蛾*（一种 19 世纪从欧洲进入美国的昆虫，其幼虫吃植物叶子），需要在大片土地上广泛而无差别地施洒杀虫剂滴滴涕（DDT）。这一项目已在森林地区造成大面积树叶脱落*现象，从而对鸟类的健康和行为造成了明显影响，鸟类被迫从无叶的树上迁离。[4]

卡森对杀虫剂（尤其是滴滴涕）的关注起源于纽约长岛地区的土地所有者与她签约调查杀虫剂一事。这些土地所有者对联邦政府在他们的土地上无差别地施洒滴滴涕表示担忧。1958年，一封写给《波士顿先驱报》的信引起了卡森的注意。信中概述了因喷洒滴滴涕灭蚊而导致的鸟类死亡。卡森在这一方面早已拥有丰富的知识储备，于是在奥杜邦鸟类协会哥伦比亚特区华盛顿分会等组织的鼓励下，她开展了进一步研究。奥杜邦鸟类协会是一个倡导保护生物的重要组织，非常支持反对化学杀虫剂的运动。[5] 协会委托卡森分析政府的喷洒计划及其后果，以帮助公众认知相关的潜在危险。

卡森也接受了一批应用型和学术型科学家的指导，后者还审阅了《寂静的春天》中的许多技术细节。她在国家健康研究院医学图书馆完成了大部分实验，并和那里的很多研究人员都有合作。其中最重要的指导来自威廉·休珀博士[*]，他是指认致癌杀虫剂的领军人物，而当时杀虫剂致癌还是一个备受质疑和争议的话题。卡森任职于美国渔业局期间，与许多政府科学家有过合作，卡森也得到了他们的支持。她很快就意识到在有关化学杀虫剂的影响上学术圈和科学圈内部观点不一，聚讼纷纭。这种纷争在《寂静的春天》出版后变得更加激烈。[6]

---

1. 罗宾·麦凯："蕾切尔·卡森与《寂静的春天》的成就"，《卫报》，2012年5月26日。

2. 麦克斯·奥尔施拉格:"艾默生、梭罗及哈德逊河学派",《改造自然》,国家人文中心,登录日期 2015 年 12 月 30 日,http://nationalhumanitiescenter.org/tserve/nattrans/ntwilderness/essays/perserve.htm。
3. 道格拉斯·布林克利:《荒野勇士:西奥多·罗斯福与美国改革运动》,纽约:哈珀永久出版社,2010 年,第 76—80 页。
4. 劳伦特·贝尔西:"吉普赛飞蛾重返东北部——十年来最糟糕的蛾灾在东北部爆发,昆虫学家束手无策",《基督教科学箴言报》,1990 年 7 月 2 日。
5. 帕特里卡·H. 海因斯:"未竟的事业——蕾切尔·卡森出版《寂静的春天》控诉滴滴涕 30 周年之际杀虫剂还在威胁人类生命",《洛杉矶时报》,1992 年 9 月 10 日。
6. 布赖恩·沃尔什:"《寂静的春天》怎样成为打响环境战争的第一枪",《时代周刊》,2012 年 9 月 25 日。

3 主导命题

要点 🗝

- 《寂静的春天》关注因使用化学杀虫剂来消灭害虫以保护农业生产而对动植物和人类造成的有害影响。
- 环境致癌领域的研究先驱威廉·休珀博士是卡森最重要的合作伙伴之一。
- 卡森站在道德立场上富有激情地申明：个人和政策制定者均有必要约束大企业，确保自然界得到保护。

核心问题

卡森《寂静的春天》的核心内容是关于农业生产中广泛而无差别地使用化学杀虫剂所造成的危害——这种观点在当时还极具争议。她也探讨了过度使用杀虫剂以及企业鼓励使用家用杀虫剂的问题。这部著作主要致力于整合不同的科学数据，以证明化学品对动植物和人类造成了各种危害。

卡森展示的数据和影响至关重要。原因如下：首先，杀虫剂的使用在当时主要被视为一种技术进步和一种让农民在提高产量的同时防止虫媒疾病*传播的必要工具。其次，政府支持杀虫剂的使用，并和大型农业公司合作，异口同声地告诉美国大众杀虫剂是安全、有益的。政府还负责在公共土地上无差别使用杀虫剂。再次，卡森所展示的研究成果使人们开始质疑大型家长式企业和政府行为的可信度。前者可以左右其客户的环境解决方案，而后者没有把民众的健康、安全、福祉放在首位。最后，《寂静的春天》是其后数十年环境监管的催化剂。

> "'控制自然'是一种傲慢的措辞,是生物学和哲学尚处于低幼阶段的产物,当时人们还认为自然界仅为服务他们而存在。这些应用昆虫学(研究昆虫的学问)的概念和做法在很大程度上应归咎于科学上的蒙昧。一种如此原始的科学竟然为自己装备了最现代、最可怕的武器,这些武器在被用来对付昆虫之余,也已成为对整个地球的威胁,这真是我们的巨大不幸。"
>
> —— 蕾切尔·卡森:《寂静的春天》

参与者

卡森非常清楚,《寂静的春天》的出版将会引起轩然大波。[1] 这部作品是对大公司所获得的权力和信任的严厉批评,也是对政府部门既支持和补贴农户又监管食品及农业生产所造成的不可避免的利益冲突的严厉批评。20 世纪 50 年代,美国的企业经常被认为是仁慈友善的,公众非常信任它们——尤其是化工、汽车和其他技术领域的企业,它们被认为是推动二战后美国经济繁荣的关键因素。[2]

当然,致力于探寻和揭示现代化学品使用真相的不只卡森一人。国家健康研究院*(NIH,从事医疗研究的联邦机构)的科学家们几年来一直在收集化学杀虫剂所产生影响的有关数据。其他一些倡导环保的重要组织(如奥杜邦鸟类协会)也一直在展开相关活动,就此问题教育民众。[3] 很多人认为卡森起到了整合各方力量的作用,她将这些问题公之于世,用自己公认的典雅文笔在公众心头留下了不可磨灭的印记。

国家癌症研究所的科学家威廉·休珀博士对卡森的研究起到了关键作用。休珀是环境致癌实验室的带头人之一,他的研究工作确立了几种杀虫剂与人类及动物癌症之间的关联。1942 年,休珀出版了《职业肿瘤与相关疾病》，这是将职业危害与癌症直接联系起来并将某些工业物质标为致癌物*的最早著作之一。[4]卡森尤其感兴趣的是休珀先前受重要化工企业杜邦公司资助从事实验室研究的经历。他在发表了工业化学品对杜邦公司员工的有害影响的相关数据后从杜邦离职。[5]

最终，卡森发现科学家们对这一问题的看法分歧明显。一方面，有些科学家支持卡森的观点，认为化学和合成杀虫剂对人类、野生动植物和生态系统有显著影响。不过，这些科学家因为担心质疑政府支持的项目和大企业的利益会引发争议而一直未将自己的大部分工作展现于公众视野之中。另一方面，卡森沮丧地发现，有些科学家完全否认杀虫剂的使用所造成的危害。作为曾经的政府雇员，卡森非常清楚，任何表明杀虫剂有害的数据都可能引发严重的经济震荡。[6]

当时的争论

卡森的《寂静的春天》一直被看作最有影响力的现代环境文学作品之一。同时，它也经常提醒人们要提防在充分了解后果前就贸然行动的危险。这种思想就是现在广为人知的预防原则*。卡森的著作以及奥尔多·利奥波德的《沙乡年鉴》（一部呼吁人类更加尊重其栖居地并更加周全地为栖居地考虑的著作）是当今环境文学的奠基之作。

卡森的著作不仅促进了环境文学的形成与发展，而且还有助于激发公众采取行动，这些行动又催生了很多法律、法规和政策，后者构成了美国今天的环境法律和环境政策。[7]《寂静的春天》使美国公众得以质疑技术与经济进步给生态和公众健康造成的影响，也在民众内心注入了时刻关注这些影响的责任感。这一态度现在通常被称作"地球管理"*。

卡森也为许多其他的环境学家和社会活动家铺平了道路,尤其是为女性活动家们,包括人类学家玛格丽特·米德*和艾琳·布罗科维奇*。布罗科维奇是一名法律职员,她带头发起了美国最大的集体起诉之一——1993年因饮用水污染对旧金山太平洋燃气和电力公司提起的诉讼。[8] 很多人相信,始于1970年、每年4月22日举行的地球日*全球庆祝活动与卡森开创性的工作和她对环境运动的勇敢领导直接相关,很多地球日庆祝活动都向她的贡献表达了敬意。[9]

---

1. 琳达·李尔:《蕾切尔·卡森——自然的见证者》,纽约:麦克米伦出版社,1998年,第446—448页。
2. 比如,可参见查尔斯·埃姆斯和雷·埃姆斯的作品及其对国际商用机器公司(IBM)的营销。艾瑞克·舒乐登福雷:《查尔斯·埃姆斯和雷·埃姆斯的电影——普遍期待感》,纽约:劳特里奇出版社,2015年。
3. 维拉·L.诺伍德:"认知自然——蕾切尔·卡森与美国环境",《符号》,1987年,第740—760页。
4. 威廉·C.休珀:"职业肿瘤与相关疾病",《职业肿瘤与相关疾病》,1942年。
5. 德夫拉·戴维斯:《抗癌秘史》,纽约:基础书局,2007年,第77页。
6. 李尔:《蕾切尔·卡森》,第330—335页。
7. 加里·克罗尔:"蕾切尔·卡森的'寂静的春天'——大众媒体与现代环保主义的起源",《理解大众科学》第10卷,2001年第4期,第403—420页。
8. 斯科特·吉勒姆:《蕾切尔·卡森——环境运动的先驱》,明尼苏达州艾迪那:ABDO出版社,2011年,第91页。
9. 比如,可参见"地球日网络"或美国公共广播公司《美国历程》系列。

4 作者贡献

要点

- 在《寂静的春天》里,卡森论证了化学杀虫剂的广泛使用正在危及地球的自然平衡,也在危及人类自身。
- 卡森将描述性和诗意的写作风格与确凿的科学数据结合起来,以一种既令公众警醒又明白易懂的方式表达自己的主张。
- 虽然卡森有关杀虫剂的论断并非属她独有,但她用清晰直接的方式呈现数据、探讨问题,这为她赢得了广泛的读者。

作者目标

作为一位受过专业培养并对大自然无限热爱的科学家,卡森对无差别使用杀虫剂所带来的危害有客观认知。她就是要借助《寂静的春天》将这种危害昭示于天下。环境作家的丰富经历,再加上作为学术人员和职业科学家所受到的精深训练,塑造了她表达这一主题的独特能力。在《寂静的春天》里,卡森的首要目标是揭示政府和企业是如何在公共土地和私人土地上无差别喷洒杀虫剂的,并警示世人当心接触和使用杀虫剂所带来的危险。

卡森提出了几个十分有力又备受争议的观点。首先,她提纲挈领地展示了她和科学家同事们收集到的有关化学和合成杀虫剂的广泛使用及其对动植物及人类造成危害的具体数据。其次,她指出了美国联邦政府组织存在内在利益冲突的问题。因为,一方面它控制着有毒(或有潜在毒性)物质的检测和监管,另一方面它又管理着支持农民、农业和食品监管的项目。第三,她提出了自己最重要的观点：技术进步（尤其是与化学品应用有关的进步）所带来的短期收益很可能造成长期的有害后果，而这种害处要远远大于收益。

卡森在追求理想的道路上并不孤单。事实上，其他科学家和环保组织邀请她加入他们的工作，发挥她的能力创作出能够直抵人心、引发思考的优美散文，这比光秃秃的数据有更好的效果。卡森也为其他无数的环保活动家铺平了道路。他们虽然关注不同的主题，但都在仿效她的方法，希望能与她的成功相媲美。

> "《寂静的春天》依然是一种在我们的自满自得中迸发出的理性声音……（卡森）把我们带回一个近乎遗失在现代文明之中的基本理念：人类与自然环境相互关联，无法分开。"
>
> —— 艾尔·戈尔：《寂静的春天》序言

研究方法

卡森的文字所展现的绝不仅是原始的数据和科学事实。这些文字描绘出了一幅自然平衡的优雅画面和与此形成强烈反差的因使用合成杀虫剂所造成的恐怖景象。这为她赢得了广泛的读者，其中包括科学家、政治家和普通大众。马克·汉密尔顿·莱特尔*在其1998年著作中将卡森称为"温柔的颠覆"。[1] 在书中，他说明了卡森的方法，即用端庄淑女的公众形象精妙地平衡了她对社会中最强大力量所做的攻击和挑战。

《寂静的春天》第一章"明天的寓言"描写了持续使用杀虫剂所造成的令人震撼的环境现实："一种奇怪的枯死病悄悄蔓延到这一带……邪恶的咒语已在这个群落降临……死亡的影子无处不

在……静得出奇。鸟儿呢？它们哪儿去了？"卡森紧接着勾勒出植物、野生动物和人类身上发生的毁灭性的、令人不安的变化，但结尾时又宽慰读者，"还没有哪个群落遭受了我所描写的不幸……"不过她警告说，"上述每一种灾难都已经实实在在地在某地发生，很多现实中的群落也已遭受相当大数量的灾难。"

时代贡献

卡森十分依赖她在美国渔业局工作期间认识的一小部分科学家的研究，他们当中很多人在国家健康研究院（NIH）和国家癌症研究所（NCI）*工作。这些科学家已经记录并分析了合成杀虫剂的危害，并特别把它们归类为致癌物，这在当时是颇具争议的结论。要不是因为卡森，他们很多人的研究就不会为公众知晓了。

卡森还依赖她在之前口碑颇佳的著作中所完善的写作方法。她既能用其优雅流畅的语言使公众保持兴趣，同时又能谨慎地把科学数据和自己的原创性研究融入书中。卡森的研究合作者们有时候对卡森着力使用"更浅易"的语言感到失望。然而，对卡森而言，她写作手法的两个方面缺一不可：文学风格可以使她永葆对自然界的惊叹与好奇，而她所接受的严谨的科学训练使她得以理解自然界。这种方法使卡森今天还能为人牢记，也对当下颇具启迪意义。

---

1. 马克·汉密尔顿·莱特尔：《温柔的颠覆——蕾切尔·卡森，〈寂静的春天〉和环境运动的兴起》，纽约、牛津：牛津大学出版社，2007年。

第二部分：学术思想

5 思想主脉

要点

- 在《寂静的春天》中,卡森研究了无差别使用杀虫剂对动植物和人类生命造成的影响与危害。
- 《寂静的春天》的中心思想是:人类的技术力量以及"人可以控制自然"的假设会产生极其可怕的长期后果。
- 《寂静的春天》将科学数据及警世故事这两种反差分明的元素结合起来表达思想,其传达的整体信息既客观可信,又能引发情感共鸣。

核心主题

在《寂静的春天》中,卡森提醒大众和政策制定者要当心使用化学和合成杀虫剂所带来的风险。更广义上讲,她是在警告整个社会要当心潜在的危险,也就是社会醉心于控制自然的技术进步,却不了解这种进步的后果所带来的危险。卡森展示的证据表明化学品能够长期存在于环境中,其危害可以从一个物种传播到另一物种(这个过程叫"生物体内积累*"),甚至可以从母亲传递给孩子。

这部著作有三个主题:化学和合成杀虫剂的危险与毒性;此类化学品是如何被草率地施用于环境的;对环境威胁较小的杀虫剂替代品。《寂静的春天》引入了一种观念,即对全人类而言,"地球管理"是一种道义责任。在谈及动物因接触杀虫剂而无辜毙亡时,卡森问道:"默许这种给活着的动物造成痛苦的行径,作为人类,我们难道不是在泯灭人性吗？"[1]

卡森的结论是：建议改变政府政策和农业作业习惯，利用健康、强壮、有复原力的生态系统内固有的抵抗力——这是大自然本身赋予的。她提供了具体的策略，比如天然杀虫剂*（非化学品的虫害控制手段）、作物轮作*（在某一区域定期更换耕种的作物品种）和生物性虫害控制（通过诸如寄生虫、捕食性昆虫等自然生物控制手段管理土地景观和农业害虫的做法）等。

> "在人类历史上，每隔一段时间就会出现一本能极大地改变历史进程的书。"
> ——美国参议员欧尼斯特·格里宁，见茱莉亚·凯勒：《汤姆叔叔的小屋》和《寂静的春天》这类书的政治影响深远，成了……改变世界的艺术"

思想探究

卡森指出化学杀虫剂滴滴涕（二氯二苯三氯乙烷）的使用对动植物和人类造成了严重伤害，《寂静的春天》中有很大一部分篇幅用于列出各种研究和数据以支持这一观点。她特别指出，在不知情或未经同意的情况下，公众正以惊人的速度被不断地暴露于滴滴涕和其他化学品之中。她将使用杀虫剂造成的危害按照不同的资源体系分为不同的章节：水、土壤、植物、动物和人类。

滴滴涕是一种人造有机化合物，1874年首次合成，其杀虫特性于1939年由瑞士化学家保罗·穆勒*发现，这项发现使他于1948年获得诺贝尔奖。美国军方在第二次世界大战期间开始使用滴滴涕控制军人头上的虱子。由于战后滴滴涕过剩，而且空军飞行员也已返家，联邦政府就决定用飞机喷洒滴滴涕来控制和消灭某些昆虫物种。这些昆虫包括吉普赛飞蛾（19世纪从欧洲输入，威胁本地植物的昆虫）和火蚁（20世纪30年代从美洲其他地方输入到美国的昆虫）。同时，政府还授权农业大剂量使用杀虫剂。

卡森及科学家同行此前已经记录了喷洒滴滴涕所造成的许多副作用，其中包括：多种鸟类的蛋壳变薄，而变薄的蛋壳导致幼鸟的成活率大幅下降。卡森书名《寂静的春天》指的就是因滴滴涕和其他化学杀虫剂的副作用所导致的未来听不到鸟叫的"寂静的春天"。卡森也提供了有关滴滴涕其他有害副作用的数据，包括癌症、内分泌干扰*（我们向血液中分泌荷尔蒙的能力失调）和早产。此外，随着滴滴涕在食物链中上移，其在生物体内的浓度通过在生物体内积累的过程而增加，这一过程进一步增加了大型哺乳动物*和人类等更大更复杂的物种体内的化学毒性。

卡森还在"不必要的灾难"这一章中详细描述了滴滴涕的广泛使用情况。她写道："我们正新增……一种灾难——通过在土地上无差别喷洒化学杀虫剂……使几乎所有类型的野生生物直接受害……在人类征伐活动中的附带受害者无足轻重。"[2] 卡森列举了一些喷洒的例子。在有些案例中，作为被消灭目标的虫害并不严重，数量不高，没有进一步调查的意义。在另一些案例中，没有证据表明所用的化学品对害虫物种是有效的，或者对同一栖息地的其他物种是安全的。[3]

《寂静的春天》中最激进的一点是卡森把滴滴涕对环境的影响比作原子弹*辐射*。当时距美国军方在日本城市长崎和广岛投下原子弹炸死十万多平民还不足20年，美国民众对此记忆犹新。她在《寂静的春天》中说:"我们有理由对辐射的遗传效应感到震惊。那么,对于广泛散播到环境中的化学品造成的同样后果,我们又怎能漠不关心?"[4]

语言表述

《寂静的春天》的独特性和影响力与其诗性的优雅、信手拈来的文字以及用语言描绘图景的能力直接相关。她能够拿出一组非常专业化的数据,将其传达给不同背景的广大受众,更重要的是,帮助他们弄明白为什么有必要关心此事。

卡森在出版《寂静的春天》之前就已经是一位知名的获奖作家,曾因《我们周围的海洋》赢得1952年美国纪实类国家图书奖。她使用大量的文学手法,尤其是警世隐喻,比如"死神的特效药"、"崩溃声隆隆"和"另一条道路"之类的标题。全书以讲故事的手法写成,同样充满了警世意味。这些手法有助于她在情感层面上直达读者内心,帮助读者建立与科学问题的个人关联。对很多读者而言,这种情感关联催生了一种义务感和责任感,激励读者对她的"行动呼唤"做出强有力的回应。

卡森非常了解她的读者,善于利用他们的恐惧和对自我利益的考量。例如,为了引起郊区家庭主妇的共鸣,她详细描述了常见动物(如浣熊和松鼠)惨死在她们修剪一新的草坪上的样子。这证明她有与读者交流的诀窍,读者能对其观点产生情感上和理性上的回应。20世纪60年代,女性尚未获得与男性同样的尊重和科学认可,她们主要关心在后院玩耍的自家孩子。

1. 蕾切尔·卡森:《寂静的春天》,纽约:霍顿·米夫林·哈考特出版社,2002年,第100页。
2. 卡森,《寂静的春天》,第85页。
3. 卡森,《寂静的春天》,第85—100页。
4. 卡森,《寂静的春天》,第37页。

6 思想支脉

要点

- 《寂静的春天》中的次要思想可以归纳为：每个人都有保护环境的道义责任。
- 卡森强调了环境保护与经济利益之间的冲突。
- 《寂静的春天》也是一个具有普遍性的行动号召，鼓励每个人去要求政府公开信息，并就这些问题进行自我教育。

其他思想

在《寂静的春天》中，卡森要求美国公众思考他们的道义责任。她说他们有义务保护自然，去帮助那些无法自助者——那些如果得不到帮助就将生存在一个受污染的世界中的野生动植物、孩子以及子孙后代。"我们对赖以生存的自然界缺乏悉心关怀，子孙后代将不会原谅这一点。"[1]

《寂静的春天》的思想基础是大多数环境政策与管理讨论中常见的观念，比如经济、商业与环境保护之间的内在冲突。卡森的立场是：自然界比公司利益更值得尊重。当然，她也使用经济论据来支持自己的观点。她说生物控制不仅让每亩土地成本更低，而且如果不需要处理中毒的野生生物或有毒食品，或者不对人类健康造成影响，那么实际上还可以节省更多费用。[2]

卡森用了一整章讨论杀虫剂、遗传学*（关于基因的科学，基因是让一些性状从一代传递到下一代的生物物质）和癌症的关系。这种观念在著作出版之时颇为新颖，鉴于当时大型化工企业的经济和政治力量，也引起了很大争议。卡森的几位科学家同仁（其中最著名的是威廉·休珀博士）多年来一直坚守在环境癌症研究领域，成就引人瞩目。卡森也特别指出了联邦政府内部固有的部门利益冲突。她断言，这些因素导致了滴滴涕的广泛而无差别的滥用，而公众对此并不知情。她在著作出版后的参议院听证会上详细阐述了这一观点。

> "湖中的芦苇枯萎了，也听不到鸟儿的鸣唱。"
> —— 约翰·济慈：《无情的妖女》

思想探究

　　卡森获得的最大褒奖之一，就是人们将其《寂静的春天》比作19世纪作家哈丽叶特·比彻·斯托*1852年发表的《汤姆叔叔的小屋》。后者引发了美国的废奴运动——该运动是为了废除奴隶制，并使美国民众意识到奴隶制是一种道义上的犯罪。它是19世纪最受欢迎也最有争议的著作之一。美国政府环境保护署（EPA）说："《寂静的春天》在环境保护史上所起的作用与《汤姆叔叔的小屋》在美国废奴运动中所起的作用并无二致。"[3]

　　将地球管理置于道德语境下审视，就使其与芸芸众生都关联起来了。环境的可持续性（利用有限的自然资源生存的能力）和地球管理也就成了道义问题。在当今全球气候变化*（主导地球天气及温度的模式的长期变化）的环境下，这种思想越来越受欢迎。根据斯坦福大学心理学教授罗布·威勒的观点："如果人们站在道德的立场上看待问题，他们的想法会很不一样。在这种情况下，人们更有可能回避或拒绝冷静的成本利益分析。相反，他们会因为觉得'这是正确的事'而选择采取行动。换言之，当我们在道义驱使下去做一件事时，更有可能为其做出贡献。"[4]

卡森明白，必须先使公众沉迷于自然之神秘和美丽，然后才能期望他们采取行动保护自然。《纽约时报》一位作者写道："卡森相信人只有热爱才会保护。"[5] 虽然卡森在前期的书里已经与读者建立了这种融洽的关系，但她在处理这种情感关联时还是与使用科学数据和引证一样严肃认真。她还浓墨重彩地描述了如果忽视她的警告则可能出现的恐怖和黯淡前景。卡森向读者解释了化学杀虫剂无影无踪、无处不在、无声无息的特点，也说明了化学杀虫剂可以多年潜伏在环境中而保持稳定，毒性不减。她明确指出，政府和企业对杀虫剂的危险心知肚明，却故意对它们的影响轻描淡写或加以隐瞒。她说，没人会相信环境及其资源可以安全如初。

被忽视之处

除了对杀虫剂使用的直接评论外，《寂静的春天》也是对二战后 20 世纪 50 年代的美国文化的间接评论。这种文化的特征是：追求消费主义（花钱和购物的文化）、追求物质主义（强调物品的物质价值及其所代表的地位的文化）、大力发展工程和科技（尤其是空间探索）、强调统一性和工业生产的现代化。第二次世界大战期间，疾病预防、化学战、军工技术（包括美国国家高速公路体系的建设）等科技领域均取得了长足的进步。20 世纪 50 年代，这些先进技术中有很多被用于民用领域，这其中就包括将杀虫剂滴滴涕用于室内和花园的家庭植物上。那时，普通百姓被灌输这样一种思想：要信任政府和大企业的领导，相信他们会充分考虑普通大众的利益和美国的健康与安全。卡森要求公众行动起来对抗政府和大公司，这是 20 世纪 60 年代向反建制文化转向的先声。[6]

卡森的声望也使她在新兴的生态女性主义运动中获得一席之地。生态女性主义是个定义宽泛的术语，它将对环境问题的关注和对传统的女性主义关注结合起来，将两方面的问题均视为男性主导社会的结果。在卡森的一个著名的论断中可见其对待自然的方式具有某些女性主义的特质。她说，"'控制自然'是一种傲慢的措辞，是当生物学和哲学还处于低幼阶段时的产物，当时人们还认为自然界仅为服务他们而存在。这些应用昆虫学（研究昆虫的学问）的概念和做法在很大程度上应归咎于科学上的蒙昧。一种如此原始的科学竟然为自己装备了最现代、最可怕的武器，这些武器被用来对付昆虫之余，也已成为对整个地球的威胁，这真是我们的巨大不幸。"[7]

学者们还在争论生态女性主义到底是不是女性主义（与争取两性平等相关的政治文化运动）或生态学的一个分支。然而，卡森本人确是在少有女性接受过科学教育、有毅力对抗男性主导的庞大工业领域之时写出了该作品。她无疑是一位先驱，为环境或其他领域的女性领袖铺平了道路。

1. 蕾切尔·卡森：《寂静的春天》，纽约：霍顿·米夫林·哈考特出版社，2002年，第15页。
2. 卡森：《寂静的春天》，第159—172页。
3. 约书亚·罗斯曼："蕾切尔·卡森的博物学"，《纽约客》，2012年9月27日。
4. 罗布·威勒："环境事关道义吗？"，《纽约时报》，2015年2月27日。
5. 伊莱扎·格里斯沃尔德：《寂静的春天》怎样引发了环境运动"，《纽约时报》，2012年9月21日。
6. 格里斯沃尔德："《寂静的春天》怎样引发了环境运动"。
7. 卡森，《寂静的春天》，第297页。

7 历史成就

要点

- 《寂静的春天》实现了其目标，即让公众知晓使用杀虫剂的影响。
- 卡森通过她颇具说服力的文学风格和她的公信力，拨动了各类读者的心弦。
- 卡森的批评者歪曲了她认为有必要对杀虫剂进行监管的主张，引发了更多批评者加入对她的声讨。

观点评价

卡森的《寂静的春天》无疑改变了世界，可以说在很大程度上引发了 20 世纪的环保运动。卡森将滴滴涕及其对生物的影响带入公众视野。许多人赞美这部著作，认为它是导致美国监管和禁止滴滴涕及其他化学杀虫剂的唯一原因。不过，也有人认为这些杀虫剂当时已经很自然地不受欢迎了。美国政府环境保护署网站公开表扬卡森为创作这本书所做出的努力。多位环境领袖和民选政治家也热烈地赞扬她为美国环境立法奠定了基础。

卡森也受到了批评。事实上，很多批评特别严苛。据报道，跨国农业化工和生物技术 * 公司孟山都 * 在 1962 至 1963 年间花费了 25 000 美元创建了公共关系团队来对付卡森，同时大力赞扬自己公司生产和销售的化学品。[1]

这类宣传活动对卡森著作的批评，几乎到了歇斯底里的程度。比如，他们给卡森贴上 "共产主义者" *（苏联 * 的支持者）的标签,或者放言,以任何形式发表反对农业生产的意见都是反美行为。然而,他们这么做却无疑为卡森和她的著作带来了更多关注,所造成的轰动效应可能比没有他们诋毁的情况要长久得多。²

> "《寂静的春天》像一吨砖头落在了婚礼仪式上。"
> ——比尔·莫耶斯,《公共广播电台杂志》,2007年

当时的成就

卡森的《寂静的春天》几乎瞬间就对美国政府产生了巨大的影响。参众两院的立法机关在该书出版后的数月内安排了多场听证会。总统和内政部长都委托机构对卡森指出的杀虫剂进行进一步研究。1970年美国环境保护署成立,人们通常也将其归功于《寂静的春天》。为纪念卡森,好几个国家博物馆举办相关展览并流动展出,其中包括位于哥伦比亚特区华盛顿的史密森尼博物馆*。蕾切尔·卡森国家野生动物保护区也于1969年在卡森位于缅因州的度假别墅附近设立,现由美国鱼类及野生植物管理局管理。

卡森思想的影响远远超出了环境、化工和政治领域。1963年6月,《大众科学》杂志发表了一篇题为"如何毒死害虫,而不是自己"的文章,概述了一些危害程度较低的昆虫控制策略。1962年和1963年,广受欢迎的连环漫画《花生》的作者查尔斯·舒尔茨*在四个不同场合称卡森为"女孩子们的女英雄"。另有几个连环漫画也提出了类似的观点。1969年,流行女歌手、歌词作者琼妮·米歇尔*用其歌曲"黄色大出租车"的歌词向卡森致敬,这首歌于2002年由"数乌鸦"合唱团重新发行。这些都证明卡森的影响力经久不息。卡森的成就同样得到了古典文化和艺术界的尊敬。

她与自然界的联系，她高超的文学技巧，还有她不屈不挠的精神都受到了赞扬。[3] 此外，《寂静的春天》已在欧洲、美洲和亚洲用15种以上的语言出版。

局限性

很难评估卡森的《寂静的春天》的成败，因为她的部分主张的效果尚不明确。比如，人们普遍认为卡森的著作促使美国20世纪70年代禁止了滴滴涕的使用。但直至1985年，美国还在生产滴滴涕供出口，当年就有300多吨出口国外。另外，其实在《寂静的春天》出版之前，滴滴涕的产量和需求量就开始趋于稳定，后来随着具有抗药性*的蚊子变种的出现又有所减少。卡森对这种有抗药性昆虫变种的出现作了预警，但这并非她本人的发现。[4]

卡森呼吁人们更新观念，关注生物性虫害控制手段。虽然这种手段在获得认证的有机农场和低碳景观园内很常见，但并没有得到推广。自《寂静的春天》出版后，发达国家显然越来越关注环境问题，然而许多医疗和科学专家，甚至营养学家和健身权威，还是认为我们所吃的食物、所喝的水和所用的化妆品中有太多的合成品和化学品。

卡森的传记作者将她描述为一位安静、内敛的女性，并不为寻求公众关注而写作《寂静的春天》。然而，直到今天，她依然是20世纪最受欢迎、最有影响和最具争议的环境运动先驱者和倡导者之一。

1. 蕾切尔·卡森中心，登录日期2015年12月2日，http://www.environmentandsociety.org/。
2. 蕾切尔·卡森中心。
3. 蕾切尔·卡森中心。
4. 关于昆虫抗药性的探讨，可见蕾切尔·卡森：《寂静的春天》，纽约：霍顿·米夫林·哈考特出版社，2002年，第245—261页。

8 著作地位

> **要 点**
>
> - 在《寂静的春天》出版前,卡森已经是一位久负盛名的获奖作家。
> - 《寂静的春天》使卡森成为环保运动的倡导者和敢于挑战权威的领袖。
> - 与乳腺癌的抗争使她参加政治和环保辩论的能力受到了限制。该书出版后不久,她就去世了。

定位

卡森最受争议也最广为人知的《寂静的春天》是她的最后一部作品。在此之前,她作为环境作家的长期职业生涯已得到普遍认可。

卡森在工厂附近长大,她每天都可以看到那些高耸的烟囱对环境的影响。对自然和写作的热爱与她的这种成长环境有着内在联系。此外,她也在动物学和环境科学*方面接受了一流的教育,而那时在科学领域获得学位的女性可谓凤毛麟角。

卡森在世时,看到了她平生出版的四部作品成为畅销书并赢得多个国家奖项(第五部于她去世后的 1964 年出版)。与环境艺术家霍华德·弗雷希*合著的《海风下》初版于 1941 年,再版于 1952 年。这部作品是在高分辨率水下摄影摄像设备出现之前对美妙而神秘的海洋生物奇观的礼赞,它奠定了卡森作为一位熟稔散文化语言、文风典雅的作家的地位。

1951 年出版的《我们周围的海洋》"一夜之间成为畅销书,使卡森成为美国公共科学的代言人和国际公认的海洋权威,并奠定了她一流自然作家的声望，"蕾切尔·卡森研究所如是道。¹ 这部著作的成功依赖于卡森的语言艺术和对环境的好奇，外加融入的科学事实。这是她充分利用在政府部门工作的研究训练，与一流科技专家合作的开始。卡森最著名的传记作者之一琳达·李尔*说："卡森没有忽视给读者带来悬念与惊奇，同时也将想象力与事实及专业知识结合起来。"² 《我们周围的海洋》赢得了 1952 年美国纪实类国家图书奖。它的成功对卡森后来《寂静的春天》的影响至关重要。而且，这本书的成功为她提供了经济保障，使她可以从政府岗位上辞职，专职从事科研工作和写作。

1955 出版的《海的边缘》进一步探讨了海洋和环境的主题，书中还包含了一个使用指南。这三部曲使卡森成为一位广受欢迎、备受推崇的环境作家，她使足不出户的读者也可以了解到科学事实，欣赏到远方的风景。《寂静的春天》可以说是卡森前期环境作品的自然延伸。这部作品将她优雅的语言与专业的调查及对数据的科学解释融为一体。

> "卡森小姐……是您带来了这一切……"
> —— 美国参议员亚伯拉罕·利比科夫，
> 载阿琳·罗达·夸拉蒂耶罗著《蕾切尔·卡森传记》

整合

卡森的背景使她得以成为影响深远的环境作家。经济困难和家庭责任迫使她离开学术界，这固然是一种苦难，但也为她成为作家铺平了道路。被迫离开大学后，她开始在政府部门工作。这份工作使她接触到各种研究方法、写作要求和一群卓越的政府科学家。这一切都有助于提升她构思和出版《寂静的春天》的能力。据朋友们所言，就连与第四期乳腺癌的搏斗也给了她激情、愤怒和信念，坚定了她在《寂静的春天》中的论断。卡森 1960 年末被确诊为乳腺癌，但她没有公开这个诊断结论，因为担心有人以她的疾病影响了科学论断为由而批评她。[3]

《寂静的春天》结合了卡森先前作品中最为成功的、特征最为鲜明的元素，这些元素有机融合，共同帮助卡森完成了新的艰巨任务。卡森对环境的捍卫，还有她所坚称的"为子孙后代保护环境是我们的职责"的立场，均是她对大自然之爱的直接延伸，也根植于她儿时在家庭农场的生活经历。作为政府科学家和作家，卡森收到了来自全国各地公众的来信和电报，这些来信和电报使她警觉地意识到公众所见到的政府对环境问题的处理方式是有问题的。对自然界的热爱与敬畏，想科学地了解自然界的意愿，以及保护它的渴望——这些都在她以前的作品中有所体现——在《寂静的春天》里都融为一体了，这正是因为她切身感受到的强烈的道义责任——要把对杀虫剂使用和化学毒素的担忧公诸天下。

意义

不可否认，《寂静的春天》是卡森所创作的最重要的作品。虽然在这部作品出版时，卡森已是一位久负盛名的作家，其科学事业已经得到公众肯定，但《寂静的春天》则使她永远地成为 20 世纪声势浩大的环境运动的发起者。卡森的声望也会一直与这部作品的声望密切相关。当年著作出版之时引起的争议，今天同样还是人们争论的热点。将卡森与她《寂静的春天》中的主张截然分开几乎是不可能的。支持者们继续称赞她为环境的拯救者。他们不敢想象，如果没有她，这个世界会发展成什么样子。而反对者们还在反驳她的主张，将经济困难和数百万人因疟疾而死亡归罪于她。因为，他们宣称，她的书阻碍了非洲使用滴滴涕消灭蚊子。[4]

---

1. 蕾切尔·卡森研究所：查塔姆大学，登录日期2015年12月12日，http://www.chatham.edu/centers/rachelcarson/。
2. 琳达·李尔：《蕾切尔·卡森——自然的见证者》，纽约：麦克米伦出版社，1998年，第441—465页。
3. 琳达·李尔，蕾切尔·卡森官网，登录日期2015年12月12日，http://www.rachelcarson.org/SeaAroundUs.aspx。
4. 威廉·苏德尔：《更遥远的海岸——蕾切尔·卡森的传奇人生》，纽约：百老汇书局，2012年，第332—336页。

第三部分：学术影响

9 最初反响

要点

- 对《寂静的春天》的回应来自两大阵营：听从她的警告，支持其目标者；想方设法证明她错了，质疑其资格和动机者。
- 卡森预测到了对这本书的负面反应，但仍继续参与论争，直至书出版不足两年后因乳腺癌去世。
- 卡森的批评者的观点差异极大，其中很多观点没有事实依据。

批评

蕾切尔·卡森非常清楚，《寂静的春天》的出版很可能招致愤怒的批评，尤其是来自化工产业的批评。一家杀虫剂滴滴涕的制造商，美国维尔斯科尔化学公司，威胁要起诉她的出版商霍顿·米夫林出版公司。这家化学公司还指控她为共产主义者。[1] 其他不那么激烈的批评集中在她对科学材料的使用上。比如有人批评卡森"挑樱桃"，即精心挑选数据支撑自己的主张而忽视那些会削弱自己主张的研究，尤其明显的是忽视环保倡导组织奥杜邦鸟类协会所做的极为全面的圣诞节鸟类统计数据。奥杜邦鸟类协会是卡森的支持者之一。虽然卡森书中说鸟类的数量在减少，但奥杜邦鸟类协会的年度调查表明鸟类的总量在上升。[2]

在2012年的一篇文章中，一位批评者说卡森"滥用、歪曲、篡改了许多她引用的研究数据，这属于科研不诚信，是一种厚颜无耻的行为。"[3] 他继而拿出证据反驳卡森的三个论断：滴滴涕导致人类癌症；滴滴涕导致鸟类数量下降；滴滴涕危害海洋。[4] 农业化工产业成员毫无意外地支持这种批评。国家农业化工品协会＊首席执行官帕克·C.布林克莱写道："杀虫剂带来的收益已大大地超额补偿了它带来的所有害处。"5

在1962年9月28日《时代》周刊发表的"杀虫剂——进步的代价"一文中，另一位批评者指控卡森过于"歇斯底里"和"女人气"。6

然而，在公众眼里，卡森是广受欢迎的。据《纽约客》杂志报道，在其发表卡森作品部分章节后收到的数百封来信中，有99%是表示支持的。有几位参议员和众议员宣读了本书的部分内容，这被写进了《国会议事录》。7

> "她知道她的主张会让99%的人吃惊。"
> ——琳达·李尔，见 www.rachelcarson.org

回应

虽然卡森在书出版后不到两年的1964年即因乳腺癌去世，但患病期间她还是在几次关键场合出面为自己的著作辩护。她曾接受美国三大电视台之一的哥伦比亚广播公司（CBS）＊一档新闻节目《哥伦比亚广播公司报道》的采访。好几位卡森的传记作家都描写了她虚弱的外表对公众的影响，尤其是对她批评者的影响。2012年，在纪念《寂静的春天》出版50周年之际，卡森的传记作者之一伊莱扎·格里斯沃尔德＊在《纽约时报》发文写道："卡森谨慎的说话方式驱散了任何认为她是工于心计者或狂热者的想法。卡森病情严重，在马里兰郊区的家中接受采访拍摄期间，她得用双手托着自己的头才行。"

卡森出席了美国参议院专门委员会于 1963 年组织的有关杀虫剂的听证会。约翰·F. 肯尼迪*总统是卡森的支持者，他命令总统科学顾问委员会（一个就科学问题向总统提供建议的团体）调查联邦政府使用杀虫剂的情况。[8] 在参议院听证会上，卡森提出了一些政策解决方案，建议将化学品管制与支持和补贴工农业的机构分离开来。多年来她一直致力于促成这一变革，因为她发现政府和大企业的利益纠缠不清，这本身就构成了问题的一部分。卡森并不要求彻底禁止杀虫剂的使用，她只是想让每个人都明白，化学品正喷洒在他们的土地上，希望他们能够有意识去控制化学品的影响。[9]

冲突与共识

虽然大众文化和集体记忆将卡森放在一个特殊的位置，仍有某些批评者质疑她的成就。很多人对她在著作出版后那么短时间内就去世感到悲伤，因为他们觉得她还可以为这个世界奉献更多，还有更好的方法帮助这个世界变得更加洁净和安全。而另外一些人则继续反驳她的观点，批评其"黑暗末日"的预测，怀疑《寂静的春天》是否真的具有其支持者所宣称的影响。

伊莱扎·格里斯沃尔德指出，该书出版之时，滴滴涕的滥用已达到了峰值。卡森指出，大约七年之内，昆虫就会对某些杀虫剂产生抗药性，因为它们的寿命和繁殖期很短。虽然很多人将后来美国禁用滴滴涕归功于卡森，但也不能只归功于她，因为那时滴滴涕的功效已经受到了质疑。格里斯沃尔德也注意到，20 世纪 60 年代初就已经出现了与政府不同的声音。虽然她认可卡森是这种文化转向的先驱者，但她也承认，卡森绝不是唯一的不同声音。而其他一些学者把诸如《洁净水法案》、《洁净空气法案》及美国政府环境保护署的设立这类环境政策的胜利直接归功于卡森和《寂静的春天》。

《寂静的春天》所针对的大型化工公司对卡森进行了激烈的反击。他们组织了大量公关活动，列出它们的杀虫剂对人类健康和环境的益处，包括防控虫媒疾病、保护食品和商业作物免受虫害等。一家大型农业化工品制造商孟山都因卡森反对其生产的滴滴涕及其他化学品而大为光火，1962年便在其公司杂志上发表了《荒凉的年景》，戏仿《寂静的春天》的开篇。这篇文章表现的是一个没有杀虫剂，昆虫控制了世界，疾病在人群中恣意蔓延的虚幻世界："想象一下吧……美国将会经历完全没有杀虫剂的一年。如是，让我们来好好看看这荒凉的年景，仔细观察一下它的毁灭性作用吧。"[10] 然而，卡森的支持者反驳了这些公关回应，并提醒人们不要忘记，自20世纪70年代初滴滴涕被禁用以来，"荒凉的年景"的预言从未真正实现过。

---

1. 伊莱扎·格里斯沃尔德："《寂静的春天》怎样引发了环境运动"，《纽约时报》，2012年9月21日。
2. 罗伯特·朱布林："关于滴滴涕和《寂静的春天》的真相"，新大西洋岛网站，2012年12月27日，登录日期2016年3月4日，http://www.thenewatlantis.com/publications/the-truth-about-ddt-and-silent-spring。
3. 查尔斯·T.鲁宾：《绿色圣战》，马里兰州兰赫姆：罗曼和利特尔菲尔德出版社，1994年，第38—44页。

4. 朱布林:"关于滴滴涕和《寂静的春天》的真相"。
5. 罗鲁斯·米尔恩和玛格丽·米尔恩:"现在我们周围全是毒",《纽约时报》1962 年 9 月 23 日。
6. 布赖恩·沃尔什:"《寂静的春天》怎样成为打响环境战争的第一枪",《时代周刊》, 2012 年 9 月 25 日。
7. 米尔恩和米尔恩:"现在我们周围全是毒"。
8. "《寂静的春天》的故事",自然资源保护咨询委员会网站,登录日期 2015 年 12 月 14 日, http://www.nrdc.org/health/pesticides/hcarson.asp.
9. 格里斯沃尔德:"《寂静的春天》怎样引发了环境运动"。
10. 孟山都公司:"荒凉的年景",《孟山都杂志》, 1962 年 10 月。

10 后续争议

要点 🗝

- 《寂静的春天》改变了公众看待环境问题的方式。
- 卡森的作品至今仍在为环境政策制定者和环境辩论提供思路。
- 美国环境保护中几个最大的变化,包括几部联邦法律的制定,应部分归功于《寂静的春天》。

应用与问题

1961年至1963年在位的美国总统约翰·肯尼迪是蕾切尔·卡森的坚定支持者。《寂静的春天》出版后,他指示总统科学顾问委员会就卡森关于杀虫剂的主张开展研究工作。该委员会于1963年在报告《杀虫剂的使用》中公布了他们的调查结果——在很大程度上与卡森的研究结果是一致的。报告鼓励联邦政府在测试毒性物质及监管毒性物质排放方面发挥更积极的作用。[1]

卡森一直在起草一些政策建议。她认为这些建议可以纠正某些政府部门的问题,而正是这些问题导致了杀虫剂的使用。这部没有发表的环境政策著作对立法产生了重要影响,并被用于一系列国会听证会和肯尼迪总统委托的专门研究中。国会以此修订了好几部法律,包括《联邦杀虫剂、灭真菌剂、灭鼠剂法案》(灭真菌剂是指用来控制诸如霉菌这类真菌的化学品,灭鼠剂是指用来控制鼠类,尤其是家鼠和田鼠的化学品)和《食品、药品、化妆品法案》。这些变化使得对化学品毒性的审查更加严格,也更好地保护了公众,使他们在日常生活中避免不知情地接触化学品,从而免受其危害。

1976年，《有毒物质控制法案》*要求环境保护署确保公众远离"对健康和环境造成伤害的不合理风险"。[2] 最终，《寂静的春天》中所指认的所有化学杀虫剂要么被禁止使用，要么被严格限制使用。

有一种对《寂静的春天》的批评是：卡森没能对使用杀虫剂带来的很多益处给予应有的肯定，尤其是在根除疟疾和脑炎这类虫媒疾病的作用方面。（疟疾是因蚊子叮咬所致的一种疾病，能引起发烧，有时甚至死亡；脑炎是引起大脑发炎的一种疾病，可以因蜱虫叮咬所致。）杀虫剂也有助于提高农业产量，而农业产量的提高又更有利于农业设备的推广和有效使用，从而减少耕作、脱粒和其他机械化过程中的排放量。[3]

许多卡森的支持者都强调，她从没有像许多批评者所宣称的那样要求全面禁止杀虫剂。相反，她只是希望在确保人人都知道杀虫剂的作用和副作用的情况下，谨慎地控制杀虫剂的使用。卡森的传记作者之一威廉·苏德尔*说："卡森并不寻求结束杀虫剂的使用，她只是想结束随意的过度使用。"

> "通过卡森的视角，我们依然可以看到不受约束的人类干预所带来的影响：她普及了现代生态学。"
>
> —— 伊莱扎·格里斯沃尔德："《寂静的春天》怎样引发了环境运动"

思想流派

关于卡森，很少有中立的观点。支持者和批评者同样都趋于片面，他们要么完全正面地看待她，要么完全负面地看待她。财产与环境研究中心*（美国一个寻求用市场手段解决环境问题的组织）说："在环境运动中，卡森几乎被誉为圣徒。"[4] 在环境教育领域，她被视为一位勇敢的领袖，不惧直接对抗大公司和联邦政府的利益，同时鼓动民众和政策制定者对化学品的有害作用采取行动。

然而，有群人不断地发声反对她。他们主要利用滴滴涕被禁用的后果作为他们论点的核心。比如，竞争性企业研究所*（设立于哥伦比亚特区华盛顿，倡导自由市场的组织）说："由于一个人按响了错误的警报，今天世界上数百万人忍受着痛苦的——甚至是致命的——疟疾带来的折磨。"[5] 相反，《时代周刊》国际编辑写道："卡森并不完美——她作品的特点在于诗性的表达，也同样在于她驾驭事实的能力——不过，让她为非洲不断爆发的疟疾而负某种责任的想法是荒谬的。"[6]

当代研究

2012年举行的《寂静的春天》出版50周年纪念活动，其中包括《〈寂静的春天〉50年——蕾切尔·卡森的虚假危机》的出版，使人们重新关注卡森。《〈寂静的春天〉50年——蕾切尔·卡森的虚假危机》由三位教授编辑，三位都与重点聚焦于环境问题的美国自由意志主义*智库（自由意志主义是一种右翼政治立场，主张政府最重要的任务就是保证个体的自由）"卡托研究所"*有联系。该书试图反驳卡森的多个独创性主张。

该书就《寂静的春天》提出了几个观点。首先，断言卡森故意只关注滴滴涕的不良影响，而对它的积极效益，尤其是在控制因蚊子所致的疟疾方面的作用避而不谈。其次，指出卡森忽视奥杜邦鸟类协会有关鸟类数量的数据，该协会的数据表明很多鸟类的数量事实上在增加，而不是在减少。第三，宣称她的癌症流行数据忽略了

重要的统计因素，如人口老龄化及吸烟引起的癌症等。⁷

不过，其他科学工作继续从卡森的作品中汲取营养。美国技术与文化学者埃德蒙德·罗素*的《战争与自然》描述了化学战与家用杀虫剂之间的关系，并直接借鉴卡森著作的内容。据英国《观察家》报纸科技编辑称，"卡森的警告与当下依然高度相关——无论是在具体的滴滴涕及其同类化学品的威胁上，还是在人类面对的总体环境危险上。"⁸ 他接着又举了几个生态学例子，证明化学杀虫剂仍然存在于各种各样的野生动植物体内。

卡森提醒公众关注科学界日益严重的忧虑，并赋予他们质询、知晓和采取行动的道义责任。书出版后，公众反响热烈。成千上万的市民给他们当地的议会代表或参议员写信请求获得知情权，并要求他们采取行动。《寂静的春天》出版后几年内，有数十家环保倡议组织成立。1970年，在卡森著作等因素的推动下，美国成立了环境保护署（EPA），这是一个联邦机构，其管理者由总统直接任命。⁹ 1980年，吉米·卡特总统因卡森的著作让公众开始关注环境问题而追授卡森"总统自由勋章"，这是美国向平民授予的最高荣誉。¹⁰

---

1. 总统科学顾问委员会（PSAC）：《杀虫剂的使用》，1963年5月15日。
2. 《有毒物质控制法案》，美国法典第15卷，第53章，1976年。
3. 财产与环境研究中心："《寂静的春天》50年：重新审视蕾切尔·卡森的经典"，登录日期2015年12月14日，http://www.perc.org/blog/silent-spring-50-reexamining-rachel-carsons-classic.

4. 财产与环境研究中心:"《寂静的春天》50 年"。
5. 伊莱扎·格里斯沃尔德:"《寂静的春天》怎样引发了环境运动",《纽约时报》,2012 年 9 月 21 日。
6. 布赖恩·沃尔什:"《寂静的春天》怎样成为打响环境战争的第一枪",《时代周刊》,2012 年 9 月 25 日。
7. 罗杰·迈纳斯、皮埃尔·德罗齐斯和安德鲁·莫里斯编,《〈寂静的春天〉50 年:蕾切尔·卡森的虚假危机》,华盛顿哥伦比亚特区:卡托研究所,2012 年。
8. 罗宾·麦凯:"蕾切尔·卡森与《寂静的春天》的成就",《卫报》,2012 年 5 月 26 日。
9. 环境保护署网站,登录日期 2015 年 12 月 5 日。
10. 吉米·卡特:"总统自由勋章颁奖典礼上的发言",1980 年 6 月 9 日。线上见格哈特·彼得斯和约翰·T.伍利:《美国总统计划》,登录日期 2015 年 12 月 5 日,http://www.presidency.ucsb.edu/ws?pid=45389。

11 当代印迹

要点

- 出版 50 多年后，《寂静的春天》依然是一部重要而有争议的环境文学作品。
- 卡森的人与自然相互影响的观点，在当今科学界和决策界仍备受争议。
- 对环境思想家们而言，面对相互矛盾的、不完整的数据，要甄别出客观真理，仍然是个挑战。

地位

对于关心环境问题和环境政策的人而言，蕾切尔·卡森的《寂静的春天》当今仍然是一部重要的著作。卡森就滴滴涕等化学杀虫剂及其长期影响有具体的主张，但这些主张的可信度仍备受争议。当前，虽然有些人对她的相关科学论断和"黑暗末日"悲观论调吹毛求疵，但人们通常还是把卡森视为英雄。持敌意批评态度的《〈寂静的春天〉50 年》这本书在很大程度上试图用与卡森相左的科学数据证明她的论断是错误的、没有事实依据的，证明她忽略了那些可以削弱她论点的证据，证明她根本没有必要那样吓唬美国公众。[1]

虽然滴滴涕在 20 世纪 70 年代初就在美国被禁用了，但释放到环境中的化学杀虫剂，或者说更广义的化学品所带来的危险，今天依然是人们担心的问题。科学家、医生、政策制定者和普通大众还在继续分析这个问题，并在诸多问题上形成了自己的观点，比如，从不含双酚基丙烷*（一种化学品，被认为与新生儿缺陷有关）的塑料瓶到饮食中有多少酱油算是合理的、有多少鸟死于风力发电机等问题。所有这些讨论可归结为一个主题：人类对自然的影响具有不确定性，有时候后果不堪设想。此外，哪些人或者哪些组织有权威或权力制定规则或设定暴露和风险的等级，都是值得讨论的。或许，卡森最重要的贡献是向公众指出他们需要独立思考的理念。

> "卡森的书甚至在成书之前就是有争议的。"
> —— 威廉·苏德尔："蕾切尔·卡森并没有使数百万非洲人毙命"

互动

时任美国副总统艾尔·戈尔在为 1994 年版的《寂静的春天》作序时说："作为一位民选官员，为《寂静的春天》作序，我心怀谦恭，因为蕾切尔·卡森的这部里程碑式的著作已无可辩驳地证明：一种思想的力量远比政治家的力量更强大。"他继续称赞卡森，认为是卡森促使其意识到环境问题。艾尔·戈尔与联合国政府间气候变化专门委员会（IPCC）* 共同获得 2007 年诺贝尔和平奖，这在一定程度上应归功于他们的主流纪录大片《难以忽视的真相》*。这部电影将环境变化问题引入流行文化之中，说明了自工业革命以来，工业行为和工业发展已经改变了地球的气候模式。

卡森的支持者和批评者继续就《寂静的春天》在滴滴涕的禁用上所发挥的作用争论不休。颇具讽刺意味的是，似乎是她的批评者，而不是她的支持者，认为禁用滴滴涕是她的功劳。卡森的批评者指责她导致了数百万非洲人的死亡。他们宣称，是她的书阻碍了滴滴涕杀灭非洲的蚊子。[2] 相反，她的支持者们认为，《寂静的春天》出版之时，至少在美国，滴滴涕已经不那么受欢迎了，因为大规模喷洒行动使得蚊子慢慢形成了对滴滴涕的耐药性。2006年世界卫生组织*（联合国负责公共健康的部门）已开始重新启动滴滴涕喷洒行动，抗击非洲疟疾。卡森的传记作者威廉·苏德尔指出，非洲从未禁用过滴滴涕。

在当今正在进行的有关气候变化的争论中，卡森的观点仍然没有过时。美国的比尔·麦奇本*和艾默里·洛文斯*等环境学者坚持认为，我们必须采取行动，以减少大气中的碳含量，降低我们对化石燃料的依赖。地球工程学*（为缓和全球范围内气候变化和海平面上升造成的影响提供技术方案的工程理论领域）的批评者常常引用预防原则——因为我们永远也说不准对地球干扰的最终后果会怎样，所以最好限制对地球的干扰。同时，消费者和公民们继续要求他们购买的产品或支持的公司提供更多的信息并提升信息透明度。

持续争议

卡森最主要的反对者通常也是环境监管的反对者，典型的是那些特别关注私有财产权、资本主义*经济市场（资本主义是一种社会经济模式，在西方占主导地位并逐渐扩展到全世界，在这种体制下，贸易和工业生产的目的均是私人利益）、大企业和保守派政治的人。丝毫不令人意外的是，化工制造商继续驳斥她的观点，因为《寂静的春天》威胁到了他们的生计。德州大学阿林顿分校经济学和法学杰出教授、财产与环境研究中心高级研究员、《〈寂静的春天〉50年》的三位主要编者之一罗杰·迈纳斯教授说："卡森的'越安全越好'的标准可见于今天的预防原则中。预防原则延迟了那些造福人类及环境的高级技术的应用。"[3] 他认为，卡森的简单化的风险观似乎对美国联邦政府的《洁净空气法案》和《洁净水法案》的起草产生过影响，而这两个法案"在某些与人类健康和技术可行性并不遥远的领域设定了不可能达到的标准"。

有关环境监管、经济增长和预防原则的争论持续主导着几乎所有有关环境问题的全球性会议，包括联合国主办的气候会议，如1992年和2012年的里约地球峰会、联合国可持续发展大会（里约+20）等。论争还将继续，因为没有明确的答案，而且事实和形势也在不断变化。但不可否认的是，卡森在过去和现在都是这场对话中的重要角色。

---

1. 罗杰·迈纳斯、皮埃尔·德罗齐斯和安德鲁·莫里斯编，《〈寂静的春天〉50年：蕾切尔·卡森的虚假危机》。华盛顿哥伦比亚特区：卡托研究所，2012年。
2. 威廉·苏德尔：《更遥远的海岸——蕾切尔·卡森的传奇人生》，纽约：百老汇书局，2012年，第332—335页。
3. http://www.masterresource.org/silent-spring-at-50/silent-spring-at-50/。

12 未来展望

要点

- 《寂静的春天》表明分析人类对环境的影响非常重要。
- 卡森力图提醒公众注意工业生产活动及环境监管漏洞,这在今天依然极具现实意义。
- 对每一位关心地球能否为子孙后代提供长期可持续性支撑的人而言,无论其对卡森的具体论断接受与否,《寂静的春天》都是行动的号角。

潜力

蕾切尔·卡森的《寂静的春天》今天依然能够强有力地揭示人们对杀虫剂的使用及总体环境的忧虑。我们今天该如何应对因全球气候变化、可吸收太阳能的气体排放、能源安全(一个国家可以稳妥地获得能源或燃料的手段)等问题所带来的政治和科学挑战?卡森的方法依然可资借鉴。世界领袖们就地球工程学的极端观点展开辩论,而自然资源保护论者则提醒我们不要忘记预防原则。

这些争辩依然会常常提及《寂静的春天》。学者们也继续讨论卡森对环境监管、政府领导、女性主义、环境保护、道德(伦理行为)等诸多领域的影响。

当今,《寂静的春天》仍然是一部被高度赞扬的纪实类文学作品。《发现》杂志将其评为"有史以来最伟大的 25 部科学著作"之一。[1] 英国《卫报》将其列为"改变世界的 50 本书"之一。[2] 它也被收录在《国家评论》编订的"20 世纪 100 部最优秀纪实类作品"

之内。《时代》杂志所列的"有史以来最伟大的纪实类作品"中，它也榜上有名。高校中有关环境科学研究与环境政策的专业越来越多，其中大部分都把《寂静的春天》列为必读书目。无论一个人对卡森的主张持何种具体观点，若想理解现代环境文学、环境政策及环境文化，均有必要阅读《寂静的春天》。

或许，卡森总领其具体观点的最重要思想就是，作为一个物种，我们永远也无法完全弄明白自己对环境的影响——要是不这么想，就显得太天真了。无论算法与计算设备多么先进，人们永远无法完整地为卡森崇敬的海洋之谜以及自然界精妙的隐秘奇观建立模型，自然界的某些方面会永远不为人类所知。因而，卡森鼓励人们带着敬意和谦卑拥抱大自然之奇迹。

> "50年后的今天，环境保护辩论的两派分坐两边。一方是被认为心地柔软的科学家以及要保持自然环境秩序的人；另一方则是来自工业领域的强硬的实用主义者（或现实主义者）以及他们位高权重的朋友们，即来自权势集团的强大力量。除了用'气候变化'代替'杀虫剂'，这场争论的内容与50年前毫无二致。"
> ——威廉·苏德尔："蕾切尔·卡森并没有使数百万非洲人毙命"

未来方向

卡森的作品持续引发热烈的争论和不断前进的学术研究。《寂静的春天》出版50周年纪念活动又激发了人们的热情，一大波关注与分析纷至沓来。卡森在继续挑战着批评者的同时也激励着新一代环境领袖。无论是那些致力于纪念卡森的机构，还是批评卡森著作的人，都将支撑起未来更多的分析，并使她的作品得以延续。

为纪念她，多家非营利性机构已经成立。卡森母校查塔姆大学的蕾切尔·卡森研究所"延续着1929届校友，被誉为帮助促成现代环境运动的作家、科学家蕾切尔·卡森的遗产"。[3] 该研究所主办了数个纪念项目，以期延续卡森著作的价值观、目标和方向，支持后代的环境学者和环境领袖。类似的机构还包括"寂静的春天研究所"，这是一个"献身科学，为公众利益服务"的研究群体，[4] 由一些知名政府机构和私人基金会资助。德国慕尼黑市的路德维希·马克西米利安大学设有"蕾切尔·卡森环境与社会中心"。美国缅因州沿海的一个野生动植物保护区是以卡森的名字命名的。卡森儿时的家在维修后被辟为博物馆。由此可见，卡森仍然活在美国及国际社会的文化记忆里。

与此相反，有几个社会活动组织至今仍在不停地批评卡森的著作和主张。网站"蕾切尔错了"（rachelwaswrong.org）继续宣称卡森应对数百万非洲人因疟疾死亡负直接责任，因为《寂静的春天》呼吁对滴滴涕实行监管。该网站称："本网站关注蕾切尔·卡森《寂静的春天》中的反技术观点带来的相关危险。这些观点弥漫于现代环境文学之中，事实上已经成为世界传统智慧的一部分。"[5]

小结

今天，蕾切尔·卡森的《寂静的春天》构成了环境文学的基石，并在更广泛的环境运动中发挥重要作用。对于那些支持其观点和目标的人而言，卡森将一如既往地作为激励和领导力之源而存在，提醒我们认识自然之美、自然之神奇、自然之脆弱，要善待一切生命。对于批评者而言，她就是"社会屈服于优雅文字之力"的明证，且她的观点需要受到批驳。对每一个普通人而言，她将一直提醒我们去质疑现状，独立思考，对最要紧的问题直言不讳地表达自己的想法。

《寂静的春天》向读者发出了令人不寒而栗的警告，它提醒人们当心一个高度工业化和唯增长论的社会可能带来的危险，为读者展现了一幅极易被摧毁或丧失的自然界图景——如果人类继续不顾后果就贸然行动的话。《寂静的春天》教导读者要兼顾好奇心和科学，而不是把他们看作相互对抗的力量，要将人类和现代社会理解为自然界的一部分而不是其敌人。它教导读者既要提出问题，也要理解答案："如果在经历了长期苦苦忍受之后我们终于提出我们有知情权，如果我们由于认识提高而发现我们正在被要求承担毫无意义且令人恐怖的风险，那么我们就应该停止接受那些让我们必须在地球上填满有毒化学品的建议；我们应该环顾四周，看看有没有其他道路向我们敞开。"[6]

---

1. "有史以来最伟大的 25 部科学著作"，《发现》，2006 年 12 月 8 日。
2. "改变世界的 50 本书"，《卫报》，2010 年 1 月 27 日。
3. 蕾切尔·卡森研究所，查塔姆大学，登录日期 2015 年 12 月 12 日，http://www.chatham.edu/centers/rachelcarson/.
4. 寂静的春天研究所，登录日期 2015 年 12 月 12 日，http://www.silentspring.org/.
5. 蕾切尔错了，登录日期 2015 年 12 月 12 日，www.rachelwaswrong.org/.
6. 蕾切尔·卡森：《寂静的春天》，纽约：霍顿·米夫林·哈考特出版社，2002 年，第 277—278 页。

术语表

1. **《难以忽视的真相》**：2006 年的一部纪录片，主角是时任美国副总统艾尔·戈尔。影片展现了气候变化背后的科学原理及气候变化的潜在影响。该片获得 2007 年奥斯卡最佳纪录片奖。

2. **原子弹**：20 世纪 40 年代研制的一种杀伤性极强的核武器，战争中已两次使用，1945 年 8 月摧毁了日本城市广岛和长崎。

3. **奥杜邦鸟类协会**：一个倡导资源保护的有影响力的组织，1905 年成立于美国。

4. **生物体内积累**：以比分解更快的速度吸收和储存任一类型化学品或化合物的过程。也可以指有机体摄入污染物后，某种物质进而沿食物链上升所产生的综合影响。

5. **杀生物剂**：任何一类可以杀死有生命动植物的化学合成剂。具体到欧洲环境立法，它被定义为一种"通过化学或生物手段消灭、阻止有害生物体或者使有害生物体无害化，或对有害生物体产生控制作用的化学物质或微生物"。

6. **生物性虫害控制**：通过使用自然生物控制手段（如寄生虫、捕食性昆虫、病原体）来管理景观和控制虫害的做法，这种方法可降低对化学杀虫剂的需求。

7. **生物学**：研究有生命生物体的科学。

8. **生物技术**：利用生物体及更广义的生物物质制造对人类有用的工具。

9. **双酚基丙烷（BPA）**：常被用作食品器皿（如金属罐、塑料食品容器、塑料水瓶）内部涂层的塑料和环氧树脂中的一种化合物。有研究表明，双酚基丙烷可以渗入到含有双酚基丙烷材料的容器内的食品或液体中。摄入双酚基丙烷对幼儿和胎儿有负面作用。

10. **癌症**：一种或一类疾病，其典型特征是细胞的异常生长。

11. 资本主义：在西方社会占主导地位并逐渐扩展到全世界的社会经济模式，以私人利益为目的从事贸易和生产。

12. 致癌物：已知会导致癌症或可能导致癌症的物质。

13. 卡托研究所：一家位于美国哥伦比亚特区华盛顿，倡导自由意志主义政治的右翼智库。

14. 哥伦比亚广播公司（CBS）：美国的一家广播公司，简称 CBS。

15. 《洁净空气法案》（1970）：一部美国法律，寻求改善、强化和推进有利于防止与减轻空气污染的项目。

16. 《洁净水法案》（1972）：一部关于综合治理水质和水污染的美国法律。

17. 共产主义者：认同共产主义政治理念者。共产主义的特征是生产资料所有权归国家所有、劳动集体化以及阶级消亡。

18. 竞争性企业研究所：一家成立于 1984 年，位于美国哥伦比亚特区华盛顿的智库，倡导通过限制政府管制来推行自由意志主义经济政策。

19. （美国）自然资源保护运动（1890—1920）：一场社会政治运动，聚焦于自然资源保护，倡导保护自然景观资源，如设立国家公园和国家森林公园。

20. 作物轮作：一种农业耕作方法，在一定时间内在某一田地或菜园内轮流种植不同种类植物，益处包括提高土壤肥力、控制害虫等。

21. 滴滴涕/二氯二苯三氯乙烷（DDT）：一种具有杀虫特性的有机氯化合物，主要用于控制因蚊子引起的疟疾。虽然它可以有效消灭对动植物有害的某些生物，但对人类和环境也极为危险。

22. 树叶脱落：使树叶落光或大部分凋落，通常是因为施用杀虫剂或遭受虫害。

23. 地球日：始于 1970 年，每年 4 月 22 日举行，旨在提醒人们注意环境问题及环境保护的重要性。

24. 地球管理：美国生态学会推广的一种理念，旨在检视各种规模的社

会生态变化及生态系统的复原力，以提高人类福祉。

25. **生态女性主义**：通常指始于 20 世纪六七十年代，由女性（尤其是美国女性）推动的旨在保护环境的政治运动。特点是将环境保护倡议及主流女性主义思想与地球灵性相结合。代表人物有玛丽·戴利、苏珊·格里芬、埃伦·威利斯和蕾切尔·卡森。

26. **生态学**：一个生物学分支，研究生物群落之间及生物群落与环境之间相互作用的方式。

27. **生态系**：在一定空间内，生物与环境构成的整体。

28. **内分泌干扰（物）**：一类能干扰人体内分泌荷尔蒙自然过程的化学物质。

29. **环境**：人和动植物生存或活动的空间和条件。

30. **环境保护**：一种在明智而审慎地使用自然资源及确保后代可持续使用环境与自然资源的同时，保持生态系统的健康和复原性的做法。

31. **环境运动**：美国（现代）环境运动始于 20 世纪 60 年代，是一种政治、科学和社会运动，旨在促使人们关注环境问题，并制定改善生态环境的策略。

32. **环境保护署（EPA）**：美国政府行政部门的一个机构，主要关注环境健康、环境质量、环境监管，在联邦、州和地方政府均有分支机构。

33. **环境科学**：有关环境的科学研究，通常借助生物学、物理学、生态学等学科开展工作。

34. **女性主义**：与争取两性平等相关的政治文化运动。

35. **遗传学**：研究基因（将生物性状从一代传递给另一代的生物物质）的学问。

36. **地球工程学**：也称为气候工程学，指的是通过人工手段改变地球气候系统，以控制大气中的二氧化碳，降低气候变化所带来的预期影响。

37. **全球气候变化**：主导地球天气及温度的模式的长期变化。

38. **吉普赛飞蛾**：从欧洲进入美国的一种飞蛾，其幼虫食用喜阴植物的叶子及其他类型的树叶。

39. **工业革命**：从18世纪60年代英国纺织制造业的机械化开始，通过新的制造工艺实现技术和工业高速增长的时期。几种气候模型表明，正是在这个时期，大气中的二氧化碳含量开始以非正常速度增加。

40. **昆虫**：最具多样性的生物群体，种类超过一百万，占据已知生物的一半以上。

41. **虫媒疾病**：由昆虫携带、传播给人类的疾病。

42. **联合国政府间气候变化专门委员会（IPCC）**：联合国的一个分支机构，成立于1988年，主要目标是"将温室气体浓度稳定在一定水平，从而防止人类对气候系统的危险性干扰"。

43. **土地伦理**：奥尔多·利奥波德《沙乡年鉴》中的一个术语，呼吁人与其栖居的土地之间建立一种全新的、更加尊重而全面的关系。

44. **白血病**：产生于骨髓的癌症类型。

45. **哺乳动物**：一种脊椎动物，特征是有毛发、中耳有三块明显的听小骨、有可产奶哺乳后代的乳腺和新皮质（大脑皮层中与视觉及听觉相关的部分）。

46. **单品种种植**：大面积种植单一作物品种的标准农作方法。这可以使效率最大化，但也会使作物更容易受到昆虫或疾病的侵害。

47. **孟山都**：1901年在美国成立的农业化工及生物技术跨国公司。

48. **国家农业化学品协会**：一家成立于美国的组织，关注诸如化肥、杀虫剂等农业化工品的使用、生产及对环境的影响。

49. **国家癌症研究所（NCI）**：美国政府卫生与公共服务部的11个机构之一，成立这个研究所的目的是进一步研究癌症的病因和治疗方法，也为那些给癌症患者及受癌症影响者提供支持的人服务。

50.《国家环境保护法案》(1970):一部美国法律,倡导在决策过程中须等同考虑环境影响的理念。

51. 国家健康研究院(NIH):美国卫生与公共服务部的联邦机构,也是联邦政府内主要的医疗研究机构。

52. 天然杀虫剂:一种用于控制虫害的非化学的或"顺势疗法"的物质,经常与生物性虫害控制策略结合使用。

53. 杀虫剂:任何一种用于控制或杀死害虫的物质,目的是保护农作物,或保持人类住所的安全和洁净。

54. 预防原则:一种风险评估方法,强调面对不确定性时要谨慎,防止在没有充分了解后果的情况下贸然行动。

55. 财产与环境研究中心:1982年成立于美国的一个组织,试图用市场手段解决环境问题。

56. 辐射:通过粒子运动(通常是电磁波)释放与传播能量。

57.(对杀虫剂的)抗药性:某种特定杀虫剂或化合物对昆虫的作用减弱的现象。在特定昆虫群落中,某些种类的昆虫自然会对杀虫剂产生抗药性。释放杀虫剂将首先杀死那些最弱的害虫个体,而留下那些对化合物有抗药性的害虫,后者在竞争者减少的环境下繁衍,从而导致抗药性群体的增加。

58. 史密森尼博物馆:世界上最大的博物馆和研究中心体系,由美国联邦政府管理,成立于1846年。

59. 苏联:存在于1922年至1991年间的共产主义国家联盟,包括俄罗斯及其东欧和亚洲北半部邻邦。

60. 合成杀虫剂:人造化学品或化学品混合物,用来预防或消灭某些被认为有害的有机体,比如昆虫、杂草、菌类等。

61.《有毒物质控制法案》(1976):美国环境保护署执行的一部美国法律,如其文字所示,"通过要求某些化学物质接受测试和必要的使用限制,达到监管商业、保护人类健康与环境等目的。"

62. 美国渔业局（现美国鱼类及野生动植物管理局）：美国内政部负责保护自然环境和野生动植物的主要机构。

63. 美国鱼类及野生动植物管理局：成立之初称"美国渔业局"，是美国内政部负责保护自然环境和野生动植物的主要机构。

64. 世界卫生组织：联合国负责全球公共健康的部门。

65. 第二次世界大战（1939—1945）：轴心国（德国、意大利、日本）和同盟国（由苏联、美国、英国领导）这两大阵营之间展开的全球战争，被视为人类历史上伤亡最惨重的冲突。

66. 动物学：研究动物生命的科学。

人名表

1. 艾琳·布罗科维奇（1960年生），美国法律职员，1993年协助发起了美国最大的诉讼之一——起诉公用事业公司太平洋燃气电力公司使用六价铬。她的故事因2000年茱莉亚·罗伯茨主演的电影《艾琳·布罗科维奇》而广为人知。

2. 拉尔夫·沃尔多·爱默生（1803—1882），美国诗人和作家。

3. 霍华德·弗雷希，美国环保主义艺术家，曾任职于《巴尔的摩太阳报》。卡森在那里与他相识。

4. 阿尔伯特·"艾尔"·戈尔（1948年生），因其在气候变化领域所做的积极工作，于2007年被授予诺贝尔和平奖。比尔·克林顿任职美国总统期间，戈尔任美国第45任副总统（任期为1993年至2001年）。

5. 伊莱扎·格里斯沃尔德（1973年生），著述广泛的美国记者及诗人，2012年《纽约时报》《〈寂静的春天〉怎样引发了环境运动》一文的作者。

6. 威廉·卡尔·休珀（1894—1978），1938—1964年任美国国家癌症研究所环境致癌部首任主任。他是环境科学家及作家蕾切尔·卡森的榜样，并为卡森写作《寂静的春天》与她进行合作研究。

7. 约翰·F.肯尼迪（1917—1963），民主党成员，1961年至1963年任美国第35任总统。

8. 琳达·李尔（1940年生），美国作家，1997年出版《蕾切尔·卡森——自然的见证者》。

9. 奥尔多·利奥波德（1887—1948），美国作家和环保主义者，因《沙乡年鉴》而闻名。《沙乡年鉴》呼吁人类与其所居住的土地之间建立一种更加尊重而全面的关系。

10. 艾默里·洛文斯（1947年生），美国物理学家和环保活动家，目前是洛基山研究所（一家环境研究咨询组织）首席科学家。他被《时代》杂志列入"2009年最具影响力的人"名单。

11. 马克·汉密尔顿·莱特尔,美国纽约州巴德学院的历史与环境研究教授。他的一部作品《温柔的颠覆》(1988)对蕾切尔·卡森的贡献做出了评价。

12. 比尔·麦奇本(1960年生),美国作家和环境保护主义者,参与创立了反对碳排放的"350.org"网站,致力于将大气中二氧化碳的浓度降至百万分之350以下。

13. 玛格丽特·米德(1901—1978),美国人类学家和作家,于20世纪六七十年代关注环境问题。

14. 琼妮·米歇尔(1943年生),加拿大歌手、歌词作者。于20世纪60年代,作为民谣歌手出道,后于20世纪70年代转向更加实验性的作品,作品取材于爵士乐和非洲音乐的传统形式。

15. 约翰·缪尔(1838—1914),苏格兰裔美国博物学家,在推动诸如约塞米蒂山谷及加州红杉国家公园等荒野地区的保护方面起到了至关重要的作用。他是环境文学传统中早期极具影响力的作家。

16. 保罗·"保利"·穆勒(1899—1965),瑞士化学家,因1939年发明杀虫剂滴滴涕而闻名,并因此获得了1948年生理学或医学诺贝尔奖。

17. 西奥多·"泰迪"·罗斯福(1858—1919),共和党成员,1901年至1909年任美国第26任总统。

18. 埃德蒙德·罗素,弗吉尼亚大学工程与应用科学学院技术、文化与传播系助理教授。

19. 查尔斯·舒尔茨(1922—2000),美国获奖漫画家,因创作漫画《花生》中的人物而闻名于世。

20. 威廉·苏德尔,美国传记作家,《更遥远的海岸——蕾切尔·卡森的传奇人生》(2013)的作者。

21. 哈丽叶特·比切·斯托(1811—1896),美国作家、反奴隶制活动家,因其《汤姆叔叔的小屋》(1852)而闻名,这部小说使全世界数百万人认识到美国非裔奴隶的悲惨境遇。

22. 亨利·大卫·梭罗(1817—1862),美国政治理论家、作家、活动家和诗人,因创造"非暴力抵抗"一词和著作《瓦尔登湖》(1854)——一部现代环境文学传统的早期作品——而闻名于世。

WAYS IN TO THE TEXT

KEY POINTS

- Rachel Carson was one of the most influential environmental* writers of the twentieth century.
- Her book *Silent Spring* is widely credited as sparking the environmental movement* in the United States.
- More than 50 years after its publication in 1962, *Silent Spring* remains a foundational text for environmental students and professionals.

Who Was Rachel Carson?

Rachel Carson (1907–64) was an American environmental writer whose greatly influential book *Silent Spring* (1962) is commonly credited with launching the environmental movement (a political and social movement emphasizing the importance of the conservation of the natural world, respecting the finite nature of our natural resources, and so on) in America in the 1960s and 1970s.

Carson's mother, a strong influence, encouraged her to develop a deep connection, respect, and love for the natural world. After a rural upbringing on the family farm in the state of Pennsylvania, she won a scholarship to the Pennsylvania College for Women, where she initially majored in English. She changed courses early in her university career, however, and graduated in 1929 with a degree in biology*—the study of living organisms. She continued her studies at the Woods Hole Institute and Johns Hopkins University, where she was awarded an MA in zoology* (the study of animal life) in 1932.[1] Carson worked as a scientist and writer for the US government for 15 years before she became an independent full-time writer.

Carson's writings, and *Silent Spring* in particular, were not only central to the launch of the environmental movement, but they had a particular influence on public awareness about the need for a more stringent regulation of pesticides*—chemicals used to control pests, notably insects,* that threaten the health of cultivated plants. The agriculture and pesticide industries attacked *Silent Spring*'s ideas furiously. In response, Carson pointed out how political loopholes allowed these industries to operate with little regard for the environmental consequences of their products and methods. She was not able to continue this work for long, however; Carson died of breast cancer in 1964 aged just 56.

What Does *Silent Spring* Say?

Silent Spring is a cautionary tale about the world and its ecosystems*—its various systems of habitats, plants, and animals—for anyone who wishes to be a thoughtful, knowledgeable citizen of the earth. It has helped millions of academics, policymakers, and concerned citizens recognize the impact of humans on nature and consider how to safeguard current and future generations by finding a sustainable way of life.

The main subject of Carson's *Silent Spring* is the negative, widespread, and long-lasting effects of human activity on the environment, which she illustrates through one major case study: the use of chemical pesticides in agriculture. Carson details the harmful and devastating results of the widespread use of synthetic chemical pesticides by farmers in the United States after World War II* (1939–45). These include everything from cancer* and birth defects in humans to the killing of small animals and birds. In fact, the "silent" of the title refers to the lack of singing birds after pesticides have destroyed the world's wildlife. Carson called for such chemicals to be renamed "biocides"*—a more accurate reflection, she believed, of their true strength and danger. The warnings of *Silent Spring* are generally credited with leading to the banning of DDT,* the pesticide dichloro-diphenyl-trichloroethane, in the United States in 1972.

Carson notes the irony that the United States government subsidized farmers to produce a surplus while claiming that chemical pesticides were needed to ensure an adequate food supply. She also criticized the political system for creating a situation that favored the claims of industry giants over the research of independent scientists and doctors into the safety of chemicals and their effects on the health of both humans and wildlife.

Silent Spring ultimately advocates the use of biological pest control.* Carson suggests using the natural checks and balances of ecosystems to control pests—for instance, diversifying crops (expanding the varieties grown) rather than planting monocultures* (growing a single species over a wide area). This, she believed, would ensure the security of the food chain while preventing damage from synthetic pesticides.* Carson's ideas about a more environmentally friendly manner of controlling agricultural insects was important, as it was a direct denial of the claims made by the chemical industries that pollution was an unavoidable consequence of an abundant and profitable food supply. She also called for a radically different approach to the use of pesticides, from one she deemed "indiscriminate" to one that carefully limits the amount and range of application.

Underpinning Carson's research on the harmful impact of chemical pesticides is a larger question: Should we try to control the environment and its ecosystems for our own, often short-term, purposes, regardless of the long-term impact on wildlife, nature, the health of the planet, or the safety and welfare of future generations?

Why Does *Silent Spring* Matter?

Silent Spring is often cited as the most important piece of modern environmental literature. It was pivotal in alerting the public to the dangers of pesticide use and to the risk of releasing into the environment chemicals whose long-term effects were not fully, or even marginally, understood. It had an equally significant impact within the scientific community.[2]

The publication of *Silent Spring* changed environmental regulation in the United States. In addition to the ultimate ban on DDT in agriculture, Carson's influence was also evident in the creation in 1970 of the Environmental Protection Agency (EPA).* This helped to separate the regulation of pesticides and chemicals from agricultural subsidies, including within the large-scale industrial operations commonly referred to as "big agro." Before this, both functions had been handled by the United States Department of Agriculture, but Carson considered this a conflict of interest.

Carson and her work continue to influence current-day policies and politicians. In the 1994 preface to *Silent Spring* written by Al Gore,* who was vice president of the United States at the time, Gore remarks that a portrait of Carson hung in his office to serve as a constant reminder of her work and its importance.³

Carson's impact reached far beyond the use of pesticides. It called into question the trustworthiness of the relationships between major corporations and government regulators, and the credibility of the sweeping claims often presented to the public as factual information but unsubstantiated by independent research. For many in the comparative comfort of postwar society, the realities presented in *Silent Spring* came as a shock. The dangers the book presented were a far cry from the marketing images of happy children frolicking on a perfectly manicured lawn found on the front of the fertilizer bag purchased at the local hardware store. If Carson's claims were true, what other horrors lay in plain sight?

Carson also helped to pave the way for other female scientists and is a forerunner of the ecofeminism* movement, which began shortly after *Silent Spring*'s publication. Ecofeminism is a political movement in which issues of environmental conservation* are combined with the principles informing the struggle for equality between the sexes. Carson remains a role model today for anyone concerned about the environment and is a constant reminder of the power of a single voice against even the most powerful of interests.

---

1. Linda Lear, *Rachel Carson: Witness for Nature* (New York: Macmillan, 1998), 72.
2. Joshua Rothman, "Rachel Carson's Natural Histories," *New Yorker*, September 27, 2012.
3. Al Gore, Introduction to *Silent Spring*, by Rachel Carson (New York: Houghton Mifflin, 1994), XVIII.

SECTION 1
INFLUENCES

MODULE 1
THE AUTHOR AND THE HISTORICAL CONTEXT

KEY POINTS

* Rachel Carson was deeply influenced by her rural upbringing on a family farm, which led to her profound understanding of the natural world.

* Carson's studies of both English and biology* enabled her to communicate complex scientific ideas clearly to a wide general readership.

* *Silent Spring* was one of the first pieces of environmental literature to be embraced by politicians and the general public when it was published in the United States in 1962.

Why Read This Text?

In the field of environmental literature, Rachel Carson is one of the most notable and quoted authors of the twentieth century; her book *Silent Spring* (1962) is required reading for all students of ecology* (the branch of biology that focuses on the relationship between groups of living things and their environment), whether or not they agree with Carson's conclusions.

Although most of the scientific data and research presented in the book concentrates on the effects of agricultural pesticides,* its themes and lessons can be applied much more generally. Indeed, *Silent Spring* is just as relevant today as it was when it was published more than 50 years ago. As a result, the number of books and scholarly articles written directly about Carson and *Silent*

Spring continues to rise.

Silent Spring also remains a prime example of the power of one person, and one voice, standing against a worldwide problem. Carson, who died soon after *Silent Spring*'s publication, continues to show how an "everyday" person can improve society and the planet. She was a pioneer in the field of modern environmental literature and a catalyst of the modern environmental movement* in the United States and elsewhere.

> *"We must not be selfish or timid if we hope to have a decent world for our children and grandchildren."*
> —— Former United States President Jimmy Carter, televised speech, April 18, 1977

Author's Life

Rachel Carson (1907–64) had a lifelong relationship with the natural world, nurtured during her childhood on her family's 65-acre farm in the district of Springdale, Pennsylvania. She was an avid reader and a skilled writer, and had her first story published in the children's magazine *St. Nicholas* at the age of 11. Most of her reading and writing focused on themes from nature and she had a particular interest in the oceans, a topic that she would write about professionally later in life.[1]

Carson excelled academically and graduated top of her high school class in 1925 before enrolling in the Pennsylvania College for Women (now called Chatham University) in Pittsburgh. She began by studying English, but soon switched to biology, though she continued to write for the university's newspaper and literary magazine. Carson completed her undergraduate program *magna cum laude* ("with great distinction") in 1929, and went on to take a master's in zoology* in 1932 at Johns Hopkins University in Baltimore, Maryland.[2]

She had begun work at Johns Hopkins on a doctoral degree, but was forced to leave in 1934 and find a job for financial reasons. Soon after, Carson's father died, and Rachel took on the full financial responsibility of caring for her mother. At the urging of a college mentor, she applied for a temporary position at the US Bureau of Fisheries* (now the US Fish and Wildlife Service, the main conservation agency within the Department of the Interior), where her primary responsibility was writing copy for radio programs. She also began to write a regular column on natural history topics for the *Baltimore Sun* newspaper. Her first notable professional publication was in July 1937, when her essay "Undersea," which beautifully described the wonders of the ocean floor, appeared in the *Atlantic Monthly*, a major American magazine. In fact, the essay was so impressive that Carson received invitations from publishers to expand it into a book.[3]

Carson had a successful career and became chief editor of publications for the US Bureau of Fisheries by 1949. She continued, though, to write and publish independently. Her first three full-length books, *Under the Seawind* (1941), *The Sea Around Us* (1951), and *The Edge of the Sea* (1955), all explored her interest in nature generally and in the oceans and aquatic life in particular.

Author's Background

The most important early influences on Carson's life are all rooted in her rural upbringing on the family farm, which gave her a fundamental love of and respect for the natural world. She understood that nature operated in mysterious, complex ways, and that it was vital to try to understand the natural world on its own terms, rather than attempt to dominate and control it for human use.

What started as a temporary position at the US Bureau of Fisheries, taken to support her family when bereavement forced her to quit the academic world, turned into a 15-year government career for Carson. The success of the books she published on environmental topics—*The Sea Around Us* and *The Edge of the Sea*—provided her with the financial means to resign from government work and devote her time and attention fully to writing.

---

1. Linda Lear, *Rachel Carson: Witness for Nature* (New York: Macmillan, 1998), 120.
2. Lear, *Rachel Carson*, 63.
3. Lear, *Rachel Carson*, 88.

MODULE 2
ACADEMIC CONTEXT

KEY POINTS

* In *Silent Spring*, Carson wrote about the dangers of widespread chemical pesticide* use in the United States.
* Building on the emerging field of environmental* literature, *Silent Spring* is considered, along with the author and environmentalist Aldo Leopold's* *Sand County Almanac*, to form the foundation of twentieth-century environmental literature.
* Carson drew upon both her rural background and her scientific training in writing the book.

The Work in Its Context

Rachel Carson's *Silent Spring* was published during a period of tremendous scientific advancement in the years following World War II.* Government and industry wholeheartedly embraced this idea of progress and fostered a culture that celebrated technology, science, and modernization. Cleanliness, standardization, technological development, quality control, and industrial mass production were all seen as signs of social progress. Human advancement and control of the natural world were interconnected themes and they were reflected in everything from putting a man on the moon to the popularization of processed, compartmentalized TV dinners.

These cultural values of scientific precision and highly technical feats of engineering also affected society's attitude toward the natural world, resulting in highly maintained, aesthetically specific landscapes (landscapes cultivated and shaped according to certain ideas of what is beautiful), the elimination of pests, and the maximization of efficiency in agricultural production. "Science and technology and those who worked in these fields were revered as the saviors of the free world and the trustees of prosperity," said Carson's biographer Linda Lear.[1]

One of the reasons behind this state of mind was the civilian use of military technology and chemicals developed during World War II, including chemical pesticides, as well as the medical advances related to the prevention and cure of insect-borne illnesses.* Another contributory factor was the rise of huge corporations in American business and industrial control over regulation (that is, control by industry of restrictions set by the government on business practices). Corporations were seen as wise, trustworthy, scientifically advanced entities that were the source of much of the postwar economic growth of the United States in the 1950s.

> "Even if she had not inspired a generation of activists, Carson would prevail as one of the greatest nature writers in American letters."
> —— Peter Matthiessen, "100 Most Influential People of the Century," *Time* magazine

Overview of the Field

The modern environmental movement* was born out of the massive increase in pollution started by the Industrial Revolution,* a period of intense technological and industrial growth that began in Britain in the late eighteenth century, and in the course of which Western societies turned from agricultural to industrial economies. This growth in pollution continued unchecked into the twentieth century. While technological advances allowed for unparalleled economic growth and provided many new jobs for working-class people, the resulting environmental and health impacts soon became apparent in deteriorating air quality, water quality, and overall environmental quality. It was also clear that human health suffered.

The conservation movement* in the United States brought about a political and cultural focus on the natural world and its preservation and conservation. It lasted from around 1890 to 1920 and built on the work of figures such as the writer and political activist Henry David Thoreau,* the poet Ralph Waldo Emerson,* and the Scottish environmental philosopher John Muir.*[2] It was popularized, too, by the United States President Theodore "Teddy" Roosevelt,* who served in office from 1901 until 1909. He championed the creation of the United States Forest Service (an organization concerned with the management of the nation's forests), signed into law the Antiquities Act, which gave the president the authority to create National Monuments (places considered significant for their special cultural or natural importance), and created five National Parks—land protected from exploitation on account of its particularly unique qualities.[3] The two World Wars of 1914–18 and 1939–45 and the catastrophic economic downturn of the 1920s and 1930s known as the Great Depression, however, shifted focus away from the environment. It was not until the late 1950s and 1960s that environmental leaders started to worry about factors such as economic growth, suburban development, paternalistic corporations (business organizations prescribing solutions and behavior for the nation's citizens), modernization and mechanization, and their effects on the natural world. This was certainly due in part to Carson's work.

In the 1950s, when Carson was working on *Silent Spring*, there was little environmental literature around. One important exception to this was the author and environmentalist Aldo Leopold's *Sand County Almanac: And Sketches Here and There*, published in 1949. In it, Leopold coined the term and developed the idea of a "land ethic,"* a philosophy that describes man's responsibility for a respectful relationship with nature and the landscape. *Silent Spring* itself would become the inspiration for many of the federal environmental regulations implemented during the administration of President Richard Nixon (1969–74), such as the Clean Water Act (1972),* the Clean Air Act (1970),* and the National Environmental Protection Act (1970)*—all laws passed to protect the nation's health and environment.

Academic Influences

The research ultimately presented as *Silent Spring* developed from Carson's other work on chemical pesticides. In the post-World War II era, the United States military funded a number of research and development programs for the agricultural application of chemical pesticides. One such program was the gypsy moth* eradication program in the mid-Atlantic region, which involved widespread, indiscriminate spraying of the pesticide DDT* on large swathes of land. Infestations of the gypsy moth, an insect* introduced from Europe in the nineteenth century whose larvae eat the leaves of plants, had caused widespread defoliation in forested areas; this had a measurable impact on the health and behavior of birds, which were forced to moved from now leafless ("defoliated")* trees.[4]

Carson's focus on pesticides—and specifically DDT—began when she was contacted by private landowners on Long Island, New York, concerned about this indiscriminate DDT spraying on their lands by the federal government. A 1958 letter to the editor of the *Boston Herald* caught Carson's attention. It outlined the avian (bird) fatalities resulting from DDT spraying to eradicate mosquitos. Carson was quite knowledgeable already on this subject but she was encouraged to research it further by such groups as the Washington, DC chapter of the Audubon Society,* a major conservation advocacy group that largely supported the crusade against chemical pesticides.[5] It commissioned Carson to study the government's spraying programs and its consequences to help it publicize the potential dangers involved.

Carson received guidance from practicing and academic scientists, who reviewed many of the technical details in *Silent Spring*. She conducted much of her research at the National Institutes of Health* Medical Library, and worked with many of the research fellows there. The most significant guidance came from Dr. Wilhelm Hueper,* a leader in the identification of cancer*-causing pesticides, a subject which at that time remained a questionable and controversial theory. Carson also drew on support from the numerous government scientists she had worked with during her tenure at the US Bureau of Fisheries.* She soon realized that the academic and scientific community was divided concerning its views on the impact of chemical pesticides. It was a difference of opinion that would only intensify after *Silent Spring* was published.[6]

---

1. Robin McKie, "Rachel Carson and the Legacy of *Silent Spring*," *Guardian*, May 26, 2012.
2. Max Oelschlaeger, "Emerson, Thoreau, and the Hudson River School," *Nature Transformed*, National Humanities Center. Accessed December 30, 2015, http://nationalhumanitiescenter.org/tserve/nattrans/ntwilderness/essays/preserva.htm.
3. Douglas Brinkley, *The Wilderness Warrior: Theodore Roosevelt and the Crusade for America* (New York: Harper Perennial, 2010), 76–80.
4. Laurent Belsie, "Gypsy Moths Return to Northeast: Worst Outbreak in a Decade Descends on Northeast; Entomologists Do Not Know How to Stop It," *Christian Science Monitor*, July 2, 1990.
5. Patricia H. Hynes, "Unfinished Business: *Silent Spring* on the 30th anniversary of Rachel Carson's Indictment of DDT, Pesticides Still Threaten Human Life," *Los Angeles Times*, September 10, 1992.
6. Bryan Walsh, "How *Silent Spring* Became the First Shot in the War on the Environment," *Time*, September 25, 2012.

MODULE 3
THE PROBLEM

KEY POINTS

* *Silent Spring* is a book about the harmful impacts to plants, animals, and humans caused by chemical pesticides* used to control insects and protect agricultural production.
* One of Carson's most significant partners was Dr. Wilhelm Hueper,* a pioneer in the field of environmental cancer.*
* Carson took a passionate, moralistic stance on the need for individuals and policymakers to keep big business in check and ensure protection of the natural world.

Core Question

At its core, Rachel Carson's *Silent Spring* is a book about the harmful effects of the widespread and indiscriminate use of chemical pesticides in agricultural production, an idea that was at that time extremely controversial. She also addresses the overuse, and industry encouragement of, domestic pesticide application by homeowners. Much of the book is devoted to bringing together different strands of scientific data that indicate these chemicals' wide variety of detrimental impacts on plants, animals, and people.

The data and impacts presented by Carson are important for several reasons. First, at the time, the use of pesticides was primarily viewed as technological progress and a necessary tool to enable farmers to increase their yields while preventing the spread of insect-borne illnesses.* Second, the government supported the use of pesticides and worked in conjunction with large agricultural corporations to tell the American public with one voice that pesticides were safe and beneficial. The government was also responsible for the indiscriminate use of pesticides on public land. Third, the presentation of Carson's research would call into question the trustworthiness both of large, paternalistic corporations—corporations that dictated things such as environmental* solutions to their customers—and of the government itself for not putting the health, safety, and welfare of its citizens first. Finally, *Silent Spring* was the catalyst for much of the environmental regulation that followed in the coming decades.

> "The 'control of nature' is a phrase conceived in arrogance, born of the Neanderthal age of biology* and philosophy, when it was supposed that nature exists for the convenience of man. The concepts and practices of applied entomology [the study of insects]* for the most part date from that Stone Age of science. It is our alarming misfortune that so primitive a science has armed itself with the most modern and terrible weapons, and that in turning them against the insects it has also turned them against the earth."
>
> —— Rachel Carson, *Silent Spring*

The Participants

Carson was well aware that the publication of *Silent Spring* would cause an uproar.[1] It is a serious critique of the power and trust given to large corporations and of the inevitable conflict of interest for government agencies that support and subsidize farmers while simultaneously regulating food and agricultural production. In the 1950s, American corporations were often considered benevolent. The public placed a good deal of trust in them, particularly those in the chemical, automotive, and other technological fields, as they were considered to be key to the post-World War II* economic prosperity enjoyed in the United States.²

Carson was not alone, however, in her quest to reveal the truth behind modern chemical use. Scientists at the National Institutes of Health (NIH),* a federal agency conducting medical research, had already been collecting data on the impact of chemical pesticides for several years, and some major environmental advocacy groups, including the Audubon Society,* had also been mounting campaigns to educate the public on the issue.³ Carson is seen by many as the unifying force who brought these concerns to light, capitalizing on her proven literary elegance to make an indelible impression on the public mind.

Of particular importance to Carson's research was Dr. Wilhelm Hueper, a scientist with the National Cancer Institute.* Hueper was one of the leaders of the environmental cancer laboratory and his work linked several types of pesticides to cancer in humans and animals. In 1942, Hueper published *Occupational Tumors and Allied Diseases*, one of the first medical books to directly link occupational hazards to cancer and label certain industrial substances as carcinogens* (substances capable of causing cancer).⁴ Dr. Hueper's previous experience as a scientist for research laboratories funded by the DuPont Corporation, a major chemical company, was of particular interest to Carson. He had left his DuPont job after publishing data on the harmful impacts of industrial chemicals on DuPont's own employees.[5]

Ultimately, Carson found the collective scientific community to be divided. On the one hand, some supported Carson's opinion that chemical and synthetic pesticides* had a measurable impact of the health of humans, wildlife, and ecosystems.* These scientists had been keeping much of their work hidden from public view, however, fearing that it would be highly controversial to contest government-supported programs and large corporate interests. On the other hand, Carson was dismayed to find that a number of scientists simply dismissed the dangers of pesticide use. Having worked as a government employee, though, she was aware that there could be serious economic repercussions from any data that suggested pesticides were anything other than harmless.[6]

The Contemporary Debate

Carson's work continues to be regarded as one of the most influential pieces of modern environmental literature and a constant reminder of the danger of taking action without understanding the consequences. This idea is now popularly known as the precautionary principle.* Carson's work, together with Aldo Leopold's* *Sand County Almanac* (a work calling for a more respectful and thoughtful relationship between human beings and the land they inhabit), forms the literary basis of the environmental and conservation movement* today.

Carson's work was not only important to the foundation of environmental literature, but it also helped stimulate the public activism that led to many of the laws, regulations, and policies that now form the field of environmental law and policy in the United States.[7] *Silent Spring* gave the American public permission to question the ecological and public health impacts of technological and economic advancement and instilled a sense of responsibility in people to keep such impacts in check. Such attitudes are now commonly referred to as "earth stewardship."*

Carson paved the way for many other environmental and social activists, particularly female ones, including the anthropologist Margaret Mead* and Erin Brockovich,* a law clerk who was primarily responsible for one of the largest class-action lawsuits, brought against the Pacific Gas and Electric Company in San Francisco in 1993 for its contamination of drinking water.[8] Many believe the worldwide celebration of Earth Day,* begun in 1970 and held every year on April 22, is directly related to Carson's groundbreaking work and fearless environmental leadership, and many Earth Day celebrations pay tribute to her work.[9]

---

1. Linda Lear, *Rachel Carson: Witness for Nature* (New York: Macmillan, 1998), 446–8.
2. See the work of Charles and Ray Eames, for example, and their marketing of the IBM Corporation. Eric Schuldenfrei, *The Films of Charles and Ray Eames: A Universal Sense of Expectation* (New York: Routledge, 2015).
3. Vera L. Norwood, "The Nature of Knowing: Rachel Carson and the American Environment," *Signs* (1987): 740–60.

4. Wilhelm C. Hueper, "Occupational Tumors and Allied Diseases," *Occupational Tumors and Allied Diseases* (1942).
5. Devra Davis, *The Secret History of the War on Cancer* (New York: Basic Books, 2007), 77.
6. Lear, *Rachel Carson*, 330–5.
7. Gary Kroll, "The 'Silent Springs' of Rachel Carson: Mass Media and the Origins of Modern Environmentalism," *Public Understanding of Science* 10, no. 4 (2001): 403–20.
8. Scott Gillam, *Rachel Carson: Pioneer of Environmentalism* (Edina, MN: ABDO, 2011), 91.
9. See, for example, the Earth Day Network or the PBS series *American Experience*.

MODULE 4
THE AUTHOR'S CONTRIBUTION

KEY POINTS

- In *Silent Spring*, Carson argues that the widespread use of chemical pesticides* is placing both the natural balance of the world and human beings in danger.
- Carson combined a descriptive, poetic writing style with hard scientific data to present her claims in a way that was both alarming and accessible to the public.
- Although Carson's claims about pesticides were not exclusively her own, she communicated the data and issues in a direct and lucid way that brought her a wide readership.

Author's Aims

As a trained scientist and long-time lover of nature, Rachel Carson was objectively aware of the dangers of indiscriminate pesticide use; *Silent Spring* was her means of announcing this to the world. Her extensive experience as an environmental* writer, combined with her in-depth training as an academic and professional scientist, made her uniquely able to communicate this message. Carson's primary aims in *Silent Spring* were to reveal how the government and industry were indiscriminately spraying pesticides on public and private lands; and to alert the world to the dangers of chemical pesticide use and exposure.

Carson makes several powerful, controversial arguments. First, she outlines specific data that she and her scientific colleagues have collected about the widespread use of chemical and synthetic pesticides* and the dangers they present to plant, animal, and human life. Second, she calls into question the inherent conflict of interest in the organization of the United States federal government. On the one hand, it controls the testing and regulation of toxic (or potentially toxic) substances, but on the other, it also runs the programs that support farmers, the agricultural industry, and food regulation. Third, she presents her most important argument—that the short-term gains afforded by technological advancement, particularly those related to chemical use, are liable to cause detrimental, long-term effects that more than outweigh the benefits.

Carson was not alone in her aims. In fact, other scientists and environmental conservation* groups recruited her to join their work, to make use of her ability to craft beautiful prose that would speak to the heart as well as to the mind in a way that was far more effective than bald scientific data. Carson also paved the way for countless other environmental activists who, while focused on different topics, imitated her approach in the hope of matching her success.

> "[Silent Spring] continues to be a voice of reason breaking in on our complacency ... [Carson] brought us back to a fundamental idea lost to an amazing degree in modern civilization: the interconnection of human beings and the natural environment."
> ——Al Gore,* Introduction to *Silent Spring*

Approach

Carson's words present much more than raw data and scientific fact. They paint an elegant picture of nature in balance, and contrast this with horrifying examples of the impact of synthetic pesticides. This brought her a wide-ranging readership that included scientists, politicians, and the general public. The author Mark Hamilton Lytle* called her a "gentle subversive" in his 1998 book of the same name.[1] In it, he chronicles the way she delicately balanced her ladylike, demure public presence with her aggressive challenge to some of the greatest powers in society.

Silent Spring's first chapter, "A Fable for Tomorrow," presents a haunting account of the environmental reality that would follow from the continued use of pesticides: "Then a strange blight crept over the area ... some evil spell had settled on the community ... everywhere was a shadow of death ... there was a strange stillness. The birds, for example—where had they gone?" Carson goes on to outline a number of devastating and disturbing changes in plant life, wildlife, and humans, but concludes by reassuring the reader that "no community has experienced all the misfortunes I describe ..." She cautions, though, that "every one of these disasters has actually happened somewhere, and many real communities have already suffered a substantial number of them."

Contribution in Context

Carson relied heavily on the research of a small community of scientists whom she met primarily during her work at the US Bureau of Fisheries.* Many of them were government employees working at the National Institutes of Health (NIH)* and the National Cancer Institute (NCI).* These scientists were already documenting and analyzing the harmful effects of synthetic pesticides and, in particular, classifying them as carcinogens.* It was a conclusion that was particularly controversial at the time and many of them would have kept their research silent were it not for Carson.

Carson also relied on the approach she had perfected in her previous highly acclaimed books. She kept the public engaged with her elegant and fluid language, carefully mixing in scientific data and her own original research. Carson's research partners were sometimes dismayed at her efforts to include "softer" language. However, for Carson these two sides to her approach—a literary style that captured her amazement and wonder of the natural world, and the scientific rigor of her training that allowed her to understand it—were inseparable. This approach helps Carson remain remembered and relevant today.

---

1. Mark Hamilton Lytle, *The Gentle Subversive: Rachel Carson, Silent Spring, and the Rise of the Environmental Movement* (New York, Oxford: Oxford University Press, 2007).

SECTION 2
IDEAS

MODULE 5
MAIN IDEAS

KEY POINTS

* In *Silent Spring*, Carson examines the impact of indiscriminate pesticide* use and the harmful effects it has on plant, animal, and human life.
* The central idea of *Silent Spring* is that man's technological power and presumption of control over nature has dire and long-term consequences.
* *Silent Spring* presents its ideas in a contrasting combination of scientific data and cautionary storytelling, which gives the overall message both objective validity and emotional connection.

Key Themes

In *Silent Spring*, Rachel Carson alerts the public and policymakers alike to the risks of chemical and synthetic pesticide* use. More broadly, she warns of the potential dangers if society commits to technological advancements leading to human control over nature without understanding the consequences. Carson shows evidence that chemicals remain in the environment* for long periods of time, and that their harmful impacts can be passed from species to species—a process called bioaccumulation*—and even from mother to child.

The work's three main themes are: the dangers and toxicity of chemical and synthetic pesticides; how such chemicals are recklessly used in the environment; and the alternatives to pesticides that pose a lesser threat to the environment. *Silent Spring* introduces the idea that earth stewardship* is a moral obligation for all human beings. Regarding the killing of innocent animals from exposure to pesticides, Carson asks, "By acquiescing in an act that causes such suffering to a living creature, who among us is not diminished as a human being?"[1]

Carson's conclusion is a recommendation that government policy and agricultural practice be changed to make use of the inherent defenses nature provides when ecosystems* are healthy, robust, and resilient. She offers specific strategies such as natural pesticides* (nonchemical means of pest control), crop rotation* (the practice of regularly changing what is cultivated in a particular place), and biological pest control* (the practice of managing landscape and agricultural pests through the use of natural biological controls such as parasites, predatory insects,* and so on).

> *"Every once in a while in the history of mankind, a book has appeared which has substantially altered the course of history."*
>
> ——United States Senator Ernest Gruening, in Julia Keller, "Works Such as *Uncle Tom's Cabin* and *Silent Spring* Had Such a Profound Political Effect that They Became ... Art that Changed the World"

Exploring the Ideas

A good portion of *Silent Spring* presents research and data that support Carson's claim that the use of DDT* (the chemical dichloro-diphenyl-trichloroethane) as a chemical pesticide causes significant harm to plants, animals, and people. In particular, she says, the public are being continually exposed to this and other chemicals at alarming rates and without their knowledge or consent. She organizes the impacts of pesticide use into separate chapters based on individual resource systems: water, soil, plants, animals, and people.

DDT is a man-made organic compound first created in 1874. Its insecticidal properties were discovered in 1939 by the Swiss chemist Paul Müller,* winning him the Nobel Prize in 1948. The US Military began using DDT to control head lice in troops during World War II,* and a surplus of DDT after the war, in addition to the fact that air force pilots were now home from combat, contributed to the federal government's decision to use it, sprayed from planes, to control and eradicate certain insect species. These included the gypsy moth* (a pest introduced from Europe in the nineteenth century that threatens native plant life) and the fire ant (an insect imported into the United States from elsewhere in the Americas in the 1930s). At the same time, the government authorized heavy use of pesticides by the agricultural industry.

Carson and her peer scientists had documented numerous harmful side effects from the spraying of DDT. These included the thinning of eggshells laid by multiple varieties of birds—and thinner eggshells lead to a significant decrease in the rate of survival of young birds. Carson's title *Silent Spring* refers to a silent future spring without birds due to the effects of DDT and other chemical pesticides. Carson also provided data on other detrimental side effects of DDT, including cancer,* endocrine disruption* (the disruption of our capacity to secrete hormones into the bloodstream), and premature birth. In addition, as it moves up the food chain, the concentration of DDT increases in organisms through a process called bioaccumulation, which increases the toxicity of the chemical for larger and more sophisticated species such as large mammals* and eventually humans.

Carson also details the widespread use of DDT in the chapter "Needless Havoc," writing that "we are adding ... a new kind of havoc—the direct killing of ... practically every form of wildlife by chemical insecticides indiscriminately sprayed on the land ... the incidental victims of [man's] crusade count as nothing."[2] Carson cites examples of this spraying. In some cases, the targeted pest was not considered to be of significant concern or population to be worth further investigation. In others, the chemicals used had not been shown to be particularly effective against the species or safe for other species sharing the same habitat.[3]

One of the most radical elements in *Silent Spring* is Carson's comparison of DDT's impact on the environment to that of radiation* from the atom bomb,* the memory of which was still prominent in the minds of the American public less than 20 years after the US military dropped atom bombs on the Japanese cities of Nagasaki and Hiroshima, killing well over 100,000 civilians. She says in *Silent Spring*, "We are rightly appalled by the genetic effects of radiation. How then, can we be indifferent to the same effect in chemicals that we disseminate widely in our environment?"[4]

Language and Expression

The uniqueness and power of *Silent Spring* is directly related to its poetic elegance, Carson's facility with words, and her ability to paint a picture with language. She is able to take a set of very technical data and communicate it to a large and diverse audience, and most importantly, help them understand why we need to be concerned.

Carson was already a well-known, award-winning author before she published *Silent Spring*, winning the National Book Award for Nonfiction in 1952 for *The Sea Around Us*. She uses numerous literary devices, particularly cautionary metaphors, with chapter titles such as "Elixirs of Death," "Rumblings of an Avalanche," and "The Other Road." The text is full of (equally cautionary) storytelling. These devices help her to reach readers on an emotional level and help them personally connect with the scientific issues. For many readers, this emotional connection generates a sense of obligation and responsibility and fuels a strong response to her "call to action."

Carson understood her audience well and played to their fears and self-interests. For instance, to strike a chord with suburban housewives, she described in detail the sight of common animals, such as raccoons and squirrels, discovered dead and disfigured on their pristine manicured lawns. This is an example of her knack of communicating with readers who would have an emotional as well as rational response to her ideas. In the 1960s, women were not afforded the same respect and scientific acknowledgement as men, but women were going to be primarily concerned for their own small children playing in the backyard.

---

1. Rachel Carson, *Silent Spring* (New York: Houghton Mifflin Harcourt, 2002), 100.
2. Carson, *Silent Spring*, 85.
3. Carson, *Silent Spring*, 85–100.
4. Carson, *Silent Spring*, 37.

MODULE 6
SECONDARY IDEAS

KEY POINTS

- The main secondary idea in *Silent Spring* is that everyone has a moral responsibility to protect the environment.*
- Carson highlights the conflicts between environmental protection and economic profit.
- *Silent Spring* is also a general call to action, encouraging everyone to demand information from government and educate themselves about these issues.

Other Ideas

In *Silent Spring*, Rachel Carson demands that the American public consider their moral obligations. They have, she says, a duty to protect nature and help those unable to help themselves—namely, wildlife, children, and future generations who will otherwise inhabit a contaminated world: "Future generations are unlikely to condone our lack of prudent concern for the integrity of the natural world that supports all life."[1]

Underlying Carson's *Silent Spring* are ideas that are common to most discussions on environmental policy and regulation, such as the inherent conflict between economics and business on the one hand, and environmental protection on the other. Carson's stance is that the natural world deserves more respect than corporate profits. However, she also uses economic arguments to her advantage. Not only can biological controls* cost less per acre, she says, but there are further savings when there are no poisoned wildlife or food products to deal with and no effects on human health.[2]

Carson devotes an entire chapter to the relationship between pesticides,* genetics* (the science of genes, the biological material allowing the transmission of characteristics from generation to generation), and cancer.* This idea was quite new at the time of publication, and considered controversial given the economic and political power of major chemical corporations. Several of Carson's scientific colleagues, most notably Dr. Wilhelm Hueper,* continued to have long and notable careers in environmental cancer research. Carson also highlights the inherent bureaucratic conflicts of interest within the federal government. These, she argues, contribute to the widespread and indiscriminate use of the pesticide DDT* without public knowledge—an idea she would elaborate on in her post-publication Senate hearings.

> "The sedge is wither'd from the lake, And no birds sing."
> ——John Keats, "La Belle Dame Sans Merci"

Exploring the Ideas

One of the most significant compliments Carson received was having *Silent Spring* compared to the nineteenth-century author Harriet Beecher Stowe's* *Uncle Tom's Cabin*. Stowe's book, published in 1852, helped galvanize the abolitionist movement in America—the movement founded in order to outlaw the institution of slavery—and made the case for the American public to consider slavery as a moral sin. It was one of the most popular and controversial books of the nineteenth century. *Silent Spring*, according to the US government's

Environmental Protection Agency (EPA),* "played in the history of environmentalism roughly the same role that *Uncle Tom's Cabin* played in the abolitionist movement."³

Putting earth stewardship* in a moral context makes it personal; environmental sustainability (the capacity to live within the means of finite natural resources) and stewardship become a moral issue. This idea is gaining popularity today in the context of global climate change* (long-term change in patterns governing the earth's weather and temperature). According to Robb Willer, Professor of Psychology at Stanford University, "People think quite differently when they are morally engaged with an issue. In such cases people are more likely to eschew [reject] a sober cost-benefit analysis, opting instead to take action because it is the right thing to do. Put simply, we're more likely to contribute to a cause when we feel ethically compelled to."⁴

Carson understood the need to first enthrall the public with nature's mysteries and wonders before expecting them to take action to protect it. According to a writer in the *New York Times*, "Carson believed that people would protect only what they loved."⁵ Having already established this rapport with readers in her previous books, Carson took this emotional connection just as seriously as she did the use of scientific data and citations. She also introduces a heavy amount of fear and gloom should her warnings not be heeded. Carson explains to her readers the invisible, pervasive, and silent quality of chemical pesticides and their ability to remain stable and potent while lying latent in the environment for many years. She is explicit that government and industry are well aware of the dangers of pesticides and are purposely downplaying or concealing their impacts. No one, she is saying, can trust that the environment or its resources will remain safe.

Overlooked

In addition to its direct commentary on pesticide use, *Silent Spring* was an indirect commentary on postwar, 1950s culture in America—one characterized by consumption (the culture of spending and shopping), materialism (a cultural emphasis on the worth of material goods and the status they confer), new advancements in engineering, science, and technology (notably space exploration), and a new focus on uniformity and factory-produced modernization. Great strides had been made in science and technology in the fields of disease prevention, chemical warfare, and military technology (including the construction of the national highway system in the United States) during World War II.* In the 1950s, many of these advancements were applied to civilian life—and this included the use of the pesticide DDT in domestic cultivation and in the house and garden. In the 1950s, the average citizen was groomed and trained to trust the leadership of the government and large corporations, who, it was believed, had the best interests of the public and the safety and security of America in mind. Carson's demand for public action against government and big business was an early forerunner of the cultural shift toward anti-establishment thought in the 1960s.[6]

Carson's reputation also earned her a place in the fledgling movement of ecofeminism,* a loosely defined term that unites environmental concerns with traditional feminist concerns, seeing both issues as resulting from male domination of society. One can see something of the feminist quality of her approach to nature in one of Carson's most famous statements: "The 'control of nature' is a phrase conceived in arrogance, born of the Neanderthal age of biology* and philosophy, when it was supposed that nature exists for the convenience of man ... It is our alarming misfortune that so primitive a science has armed itself with the most modern and terrible weapons, and that in turning them against the insects* it has also turned them against the earth."[7]

Scholars debate whether ecofeminism is a true subfield of either feminism* (the political and cultural movement associated with the struggle for equality between the sexes) or ecology.* Carson herself, though, was writing at a time when few women were educated in the sciences or had the tenacity to oppose huge, male-dominated industries. She was certainly a pioneer, then, paving the way for other women leaders in the environmental field and elsewhere.

---

1. Rachel Carson, *Silent Spring* (New York: Houghton Mifflin Harcourt, 2002), 15.
2. Carson, *Silent Spring*, 159–72.
3. Joshua Rothman, "Rachel Carson's Natural Histories," *New Yorker*, September 27, 2012.
4. Robb Willer, "Is the Environment a Moral Cause?" *New York Times*, February 27, 2015.
5. Eliza Griswold, "How 'Silent Spring' Ignited the Environmental Movement," *New York Times*, September 21, 2012.
6. Griswold, "How 'Silent Spring.'"
7. Carson, *Silent Spring,* 297.

MODULE 7
ACHIEVEMENT

KEY POINTS
- *Silent Spring* achieved its goal of informing the public about the impacts of pesticide* use.
- Carson's persuasive literary style and her established credibility allowed her to strike a chord among a diverse readership.
- Carson's critics distorted her claims concerning the need for pesticide regulation, attracting other critics to join in their condemnation.

Assessing the Argument

Rachel Carson's *Silent Spring* undoubtedly changed the world, and is largely credited with igniting the environmental movement* in the twentieth century. Carson brought DDT* and its impact on all living things into the public forum. Many credit her book as the single reason why DDT and other chemical pesticides were regulated and banned in the United States, although others believe that they would have naturally fallen out of favor. The website of the US government's Environmental Protection Agency* publicly gives Carson's efforts credit for its own creation, and many environmental leaders and elected politicians give her the overwhelming credit for setting the stage for all major environmental legislation in the United States.

Carson was not, though, without her critics—many of them particularly harsh. The multinational agricultural chemical and biotechnology* company Monsanto* reportedly spent more than

$25,000 in 1962 and 1963 to create a public relations case against her and to sing the praises of the chemicals it manufactured and sold.¹ Such campaigns were critical of Carson's work to the point of near-hysteria—labeling her a communist* (a sympathizer of the Soviet Union),* for example, or claiming that speaking against farming in any way is un-American. However, they also no doubt drew attention to her and her book and kept them both in a sensational limelight far longer than they might otherwise have been.²

> "Silent Spring *fell like a ton of bricks on a wedding party."*
> —— Bill Moyers, *PBS Journal*, 2007

Achievement in Context

Carson's *Silent Spring* had an almost immediate, measurable impact on the United States government. Many hearings in the law-making bodies of Senate and House of Representatives were scheduled in the months after the book's publication and both the president and secretary of the interior commissioned further studies on the pesticides Carson identified. The formation of the Environmental Protection Agency in 1970 is often credited to *Silent Spring*, and several national museum exhibits have been developed and circulated in Carson's honor, including one at the Smithsonian* museum in Washington, DC. The Rachel Carson National Wildlife Refuge near her vacation home in Maine was established in 1969 and is administered by the US Fish and Wildlife Service.*

Carson's ideas reached well beyond environmental, chemical,

and political circles. In June 1963, the magazine *Popular Science* published the article "How to Poison Bugs, but NOT Yourself," outlining strategies on less harmful insect* control. In 1962 and 1963, Charles Schultz,* the cartoonist behind the popular *Peanuts* comic strip, referred to Carson on four separate occasions as a "girl's heroine." Several other comic strips followed suit. In 1969, Joni Mitchell,* a popular female singer-songwriter, paid respect to Carson in the lyrics of her song "Big Yellow Taxi"; the song was rereleased by the group Counting Crows in 2002, evidence of her staying power. Carson's legacy is also honored in many classical and artistic circles, with her connection to the natural world, her superb literary skill, and her unyielding spirit being applauded.[3] Finally, *Silent Spring* has been published in more than 15 languages across Europe, the Americas, and Asia.

Limitations

It can be difficult to assess the relative success or failure of Rachel Carson's *Silent Spring*, as the outcomes of some of her demands are still unclear. For instance, Carson is widely credited with the banning of DDT in the United States in the 1970s, but it continued to be produced for export until as late as 1985, when over 300 tons were exported. Even before the publication of *Silent Spring*, however, production and demand for DDT were beginning to stabilize and later wane as resistant strains of mosquitos were beginning to emerge. Carson forewarned of the development of resistant insect strains, but this was not her own discovery.[4]

Carson called for a renewed focus on biological pest control;*

while this is now common in certified organic farms and low-carbon landscapes, it is not widely used. The developed world has certainly been more focused on environmental matters since the publication of *Silent Spring*, but many medical and scientific experts, and even nutritionists (experts in nutrition) and fitness gurus feel that there are still too many synthetic compounds and chemicals in the food we eat, the water we drink, and the cosmetics we apply.

Carson's biographers paint her as a quiet, private woman who did not write *Silent Spring* because she sought the spotlight. She nevertheless remains one of the most popular, influential, and controversial forerunners and advocates of the environmental movement of the twentieth century.

---

1. Rachel Carson Center, accessed December 2, 2015, http://www.environmentandsociety.org/.
2. Rachel Carson Center.
3. Rachel Carson Center.
4. See Rachel Carson, *Silent Spring* (New York: Houghton Mifflin Harcourt, 2002), 245–61, Chapter 15, "Nature Fights Back," for a discussion on insect resistance.

MODULE 8
PLACE IN THE AUTHOR'S WORK

KEY POINTS

- Carson was already a well-established and prize-winning author before *Silent Spring* was published.
- *Silent Spring* transformed Carson into an environmental* advocate and leader unafraid to challenge authority.
- Her ability to engage in political and environmental debate was limited, however, by her battle with breast cancer;* she died shortly after the book's publication.

Positioning

Silent Spring, by far Rachel Carson's most controversial and well-remembered major publication, was her last. She had already had a long and well-received career as an environmental writer, however.

Carson grew up near factories whose environmental impact from smokestacks was visible on a daily basis; while her deep love of nature and her writing were intrinsically linked to this upbringing, she also had a top-tier education in zoology* and environmental science* at a time when very few women achieved degrees in the sciences.

Rachel Carson lived to see the four major books she published in her lifetime (a fifth was published after her death in 1964) become best sellers and win several national awards. *Under The Sea Wind*, originally published in 1941 and republished in 1952, was a partnership with the environmental artist Howard Frech.*

It is a beautiful and mystical celebration of the wonders of life under the sea at a time before high-definition underwater cameras, photographs, and films. It established Carson as a writer at ease with literary prose and an elegant use of language.

The Sea Around Us, published in 1951, "became an overnight best seller and made Rachel Carson the voice of public science in America, an internationally recognized authority on the oceans, and established her reputation as a nature writer of first rank," says the Rachel Carson Institute.[1] This book built on Carson's use of language and her wonder about the environment but also incorporated scientific fact. This was the start of her collaboration with other leading scientific experts and put her training in government research to good use. Linda Lear,* one of Carson's most famous biographers, says, "Carson does not neglect mystery and wonder but blends imagination with fact and expert knowledge."[2] *The Sea Around Us* won the National Book Award in the United States for nonfiction in 1952, and its success was critical to Carson's later impact with *Silent Spring*. Moreover, the sales of this book provided her with the financial security to resign from her government position and pursue research and writing full time.

The Edge of the Sea, published in 1955, further built on the themes of the ocean and the environment, but also incorporated a practical user guide. This trilogy of works made Carson well known as a popular, highly regarded environmental author who made scientific fact and faraway places accessible to the household reader. *Silent Spring*, then, can be seen as a natural extension of Carson's portfolio of environmental work. It combines her graceful language with the scientific data she was qualified to investigate and interpret.

> "Miss Carson ... You are the lady that started this ..."
> —— United States Senator Abraham Ribicoff, in Arlene Rodda Quaratiello, *Rachel Carson: A Biography*

Integration

Carson's background enabled her to become an environmental writer of lasting significance. Even such hardships as the financial struggles and family obligations that prompted her to leave academia would lead to opportunities that paved the way for her role as an author. Forced to leave university, she took up a government job that exposed her to research methods, writing demands, and a network of highly skilled government scientists, all of which directly contributed to her ability to conceive and publish *Silent Spring*. Even her battle with stage IV breast cancer gave her, according to her friends, a passion, anger, and conviction that strengthened her argument in *Silent Spring*. Carson was diagnosed with breast cancer in late 1960, but kept her diagnosis hidden for fear of being criticized on the grounds that her own illness was influencing her scientific arguments.[3]

Silent Spring integrates the most successful and defining elements of Carson's previous work, brought together in the service of a new struggle. Carson's defense of the environment, together with her stance that it is our obligation to protect it for future generations, is a direct extension of her love of nature and her childhood roots on her family's farm. As a government scientist and author, Carson received letters and telegrams from citizens throughout the country alerting her to the questionable treatment of the environment they witnessed. Her love and awe for the natural world, her desire to understand it scientifically, and her yearning to protect it, all evident individually in her previous books, come together in *Silent Spring*, supported by her own strongly felt moral obligation to bring to light concerns about pesticide use and chemical toxins.

Significance

Silent Spring was undeniably the most significant piece of work Carson ever produced, and although she was already a well-known writer with an established scientific career by the time of its publication, *Silent Spring* would forever define her along with the emergence of the broader environmental movement* of the twentieth century. Carson's own reputation continues to be tied to the work's reputation, and the same controversies that emerged immediately after its publication continue to be hotly debated today. Separating Carson from her claims in *Silent Spring* is almost impossible. Supporters continue to hail her as an environmental savior and shudder to think what the world might have been like without her; her opponents, meanwhile, still refute her claims and ascribe to her economic hardship and the death of millions killed by malaria because her book, they claim, prevented DDT from being used to eradicate mosquitos in Africa.[4]

1. Rachel Carson Institute, Chatham University, accessed December 12, 2015, http://www.chatham.edu/centers/rachelcarson/.
2. Linda Lear, *Rachel Carson: Witness for Nature* (New York: Macmillan, 1998), 441–65.
3. Linda Lear, RachelCarson.org, accessed December 12, 2015, http://www.rachelcarson.org/SeaAroundUs.aspx.
4. William Souder, *On a Farther Shore: The Life and Legacy of Rachel Carson* (New York: Broadway Books, 2012), 332–6.

SECTION 3
IMPACT

MODULE 9
THE FIRST RESPONSES

KEY POINTS

* Responses to *Silent Spring* came from two camps: those who heeded Carson's warnings and supported her goals; and those who sought to prove her wrong and question her qualifications and motives.
* Carson anticipated the negative reactions to the book and remained engaged with the debate until she lost her life to breast cancer less than two years after publication.
* Carson's critics had widely varying arguments, many with little factual basis.

Criticism

Rachel Carson was well aware that the publication of *Silent Spring* would bring angry criticism, particularly from the chemical industry. One manufacturer of the pesticide* DDT,* the US chemical company Velsicol, threatened to sue her publisher, Houghton Mifflin. It also accused her of being a communist.*[1] Other less dramatic critiques focused on her use of scientific material. Carson was accused of "cherry-picking," or carefully selecting data to support her claims while ignoring research that weakened her argument. Of particular note are bird count data from the notably comprehensive Christmas Bird Count conducted by the environmental* advocacy group the Audubon Society.* Even though the Audubon Society was a supporter of Carson's, its annual survey showed overall bird populations increasing at the same time that she was writing about their decline.[2]

In an article of 2012, one critic states that Carson "abused, twisted, and distorted many of the studies that she cited, in a brazen act of scientific dishonesty."[3] He goes on to present evidence to contest three of Carson's major claims: that DDT causes cancer* in humans; that it causes bird populations to decline; and that it damages the oceans.[4] Not surprisingly, members of the agricultural chemical industry supported such claims. Parke C. Brinkley, chief executive officer of the National Agricultural Chemicals Association,* wrote, "Any harm that is caused by the use of pesticides is greatly overcompensated by the good they do."[5]

In the *Time* article "Pesticides: The Price for Progress," published on September 28, 1962, another critic accused Carson of being too "hysterical" and "feminine."[6]

In the public eye, however, Carson was largely embraced. The *New Yorker* magazine reported that 99 percent of the hundreds of letters it received in response to its publication of sections of her work were favorable, and several members of the Senate and Congress House of Representatives read excerpts into *Congressional Record*.[7]

> "She knew her claims would surprise 99 out of 100 people.'"
> —— Linda Lear, www.rachelcarson.org

Responses

Although Carson succumbed to breast cancer in 1964, less than two years after publication, she made a few key appearances to defend her work. She was interviewed on *CBS Reports*, a news program aired by CBS,* one of the three largest television stations in the United States. Several of Carson's biographers reported the impact her frail appearance had on the public, and in particular on her critics. "Carson's careful way of speaking dispelled any notions that she was a shrew or some kind of zealot. Carson was so sick during filming at home in suburban Maryland that in the course of the interview, she propped her head on her hands," wrote one of Carson's biographers, Eliza Griswold,* in the *New York Times* in her 2012 article celebrating the 50th anniversary of *Silent Spring*'s publication.

Carson made an appearance at a United States Senate subcommittee hearing on pesticides in 1963. President John F. Kennedy,* a supporter of Carson's, had already ordered the President's Science Advisory Committee (a body instituted to advise the president on scientific matters) to investigate the federal government's use of pesticide.[8] In the Senate hearing, Carson presented a number of policy solutions that sought to separate the regulation of chemicals from the agencies that supported and subsidized industry and agriculture. She had been hard at work on this change for many years, as she saw the intertwined interests of government and big business as part of the problem. Carson did not demand a complete ban on pesticides. She simply wanted everyone to know that these chemicals were being sprayed on their land and to be able to control their impact.[9]

Conflict and Consensus

Although popular culture and collective memory hold Carson

in a special place, some critics question her legacy. Many were saddened by her death so soon after the publication of *Silent Spring*, as they felt she had much more to offer the world and many more ways to make it a cleaner, safer place. Others, though, continued to refute her claims, criticize her "doom and gloom" predictions, and question whether *Silent Spring* really had the impact its supporters claimed.

Eliza Griswold notes that the rampant use of DDT was, by the time of the book's publication, starting to reach its peak. Carson herself noted that insects* develop resistance* to specific strains of pesticides within approximately seven years, because their life and reproductive cycles are so short. So while many credit Carson for the later banning of DDT, it is unlikely that she was solely responsible, as its efficacy was already in question. Griswold also notes the beginnings of dissent against the government in early-1960s America, and while she acknowledges Carson as a forerunner in this cultural shift, she admits she was by no means alone. Other scholars directly attribute significant environmental policy victories in the United States such as the Clean Water Act,* the Clean Air Act,* and the establishment of the US government's Environmental Protection Agency* to Carson and *Silent Spring*.

The major chemical companies targeted by *Silent Spring* fought back hard, creating massive public relations campaigns outlining the benefits to human and environmental health afforded by their pesticides. These included the prevention and control of insect-borne illnesses* and the protection of food and commercial crops against infestation. Monsanto,* a large agricultural chemical manufacturer angered by Carson's claims against DDT and other chemicals it produced, published "The Desolate Year," a parody of the opening chapter of *Silent Spring*, in its corporate magazine in 1962. This evoked a fantasy world with no pesticides, with insects controlling the world and disease rampant among populations: "Imagine ... the United States were to go through a single year completely without pesticides. It is under that license that we take a hard look at that desolate year, examining in some detail its devastations."[10] Carson supporters, however, refute these public relations responses, and remind us that since DDT was banned in the early 1970s, the predictions of "The Desolate Year" have never materialized.

---

1. Eliza Griswold, "How 'Silent Spring' Ignited the Environmental Movement," *New York Times*, September 21, 2012.
2. Robert Zubrin, "The Truth about DDT and *Silent Spring*," *The New Atlantis.com*, September 27, 2012, accessed March 4, 2016, http://www.thenewatlantis.com/publications/the-truth-about-ddt-and-silent-spring.
3. Charles T. Rubin, *The Green Crusade* (Lanham, MD: Rowman & Littlefield, 1994), 38–44.
4. Zubrin, "The Truth about DDT and *Silent Spring*."
5. Lorus Milne and Margery Milne, "There's Poison All Around Us Now," *New York Times*, September 23, 1962.
6. Bryan Walsh, "How *Silent Spring* Became the First Shot in the War over the Environment," *Time*, September 25, 2012.
7. Milne and Milne, "There's Poison All Around Us Now."
8. "The Story of *Silent Spring*," Natural Resources Defense Council website, accessed December 14, 2015, http://www.nrdc.org/health/pesticides/hcarson.asp.
9. Griswold, "How 'Silent Spring.'"
10. Monsanto Corporation, "The Desolate Year," *Monsanto Magazine*, October 1962.

MODULE 10
THE EVOLVING DEBATE

KEY POINTS

- *Silent Spring* changed the way the public understood environmental* concerns.
- Environmental policymakers and debates continue to draw on Carson's work.
- Some of the biggest changes in environmental protection in the United States, including several federal laws, are credited in part to *Silent Spring*.

Uses and Problems

US President John F. Kennedy,* in office from 1961 to 1963, was a great supporter of Rachel Carson, and when *Silent Spring* was released, he instructed the President's Science Advisory Committee to research her claims about pesticides.* The committee's findings were published in 1963 in the report *The Use of Pesticides*. It largely concurred with Carson's findings and encouraged the federal government to take a more aggressive role in testing toxic substances and regulating their release into the environment.[1]

Carson had been working on a number of policy recommendations she thought could rectify some of the bureaucratic issues that contributed to pesticide use. This unpublished work on environmental policy had a significant impact on legislation and was used in a series of Congressional hearings and special studies commissioned by President Kennedy. As a result, Congress amended several pieces of legislation, including the Federal Insecticide, Fungicide, and Rodenticide Act and the Food, Drug, and Cosmetic Act (fungicides are chemicals used to control fungus such as mold; rodenticides are chemicals used to control rodents, notably mice and rats). These changes increased the rigor of chemical toxicity reviews and better protected the public from the unknown presence and impact of chemicals in their daily lives. In 1976, the Toxic Substance Control Act* required the Environmental Protection Agency* to ensure public protection from the "unreasonable risk of injury to health or the environment."[2] Eventually, all the chemical pesticides identified in *Silent Spring* either were banned or their use was greatly restricted.

One criticism of *Silent Spring* is that Carson fails to give credit to the numerous beneficial uses of pesticides, particularly in eliminating insect-borne illnesses* such as malaria (a mosquito-borne disease which leads to fever and sometimes death) and encephalitis (a disease causing inflammation of the brain which can be borne by ticks). Pesticides can also help increase agricultural yield, which in turn promotes a more efficient use of agricultural equipment, reducing emissions from plowing, threshing, and other mechanized processes.[3]

Many of Carson's supporters make the point that she never called for an all-out ban on pesticides, as many of her critics claim. Instead, she wanted their prudent, controlled use while ensuring everyone knew about their use and their side effects. William Souder,* a Carson biographer, states that "Carson did not seek to end the use of pesticides—only their heedless overuse."

> "We still see the effects of unfettered human intervention through Carson's eyes: she popularized modern ecology.*"
> —— Eliza Griswold,* "How 'Silent Spring' Ignited the Environmental Movement"

Schools of Thought

There are few neutral opinions about Rachel Carson. Supporters and critics alike are passionate and tend to hold her in either an entirely negative or an entirely positive regard. The Property and Environment Research Center,* an organization in the United States seeking market solutions to environmental problems, says "Rachel Carson is hailed as a near saint in the environmental movement."*4 In the field of environmental education she is regarded as a brave leader, unafraid to take on the interests of big business or the federal government while rousing the public and policymakers to action about the harmful effects of chemical pesticides.

A significant group, however, continues to speak out against her, primarily using the banning of DDT as the crux of its argument. For example, the Competitive Enterprise Institute,* a free-market advocacy group based in Washington, DC states, "Today, millions of people around the world suffer the painful and often deadly effects of malaria because one person sounded a false alarm."5 In contrast, *Time* magazine's foreign editor has written, "Carson wasn't perfect—the quality of her book is as much in its poetry as in her ability to marshal facts—but the notion that she is somehow responsible for the continued scourge of malaria in

Africa is absurd."⁶

In Current Scholarship

Silent Spring celebrated its 50th anniversary in 2012, bringing Carson much renewed attention, including through the publication of *Silent Spring at 50: The False Crises of Rachel Carson*. Edited by a team of three professors all with ties to the Cato Institute,* an American libertarian* think tank with a strong focus on environmental issues, it seeks to refute many of her original claims (libertarianism is a right-wing political position according to which a government's most important task is to guarantee the liberty of the individual).

The book makes several points about *Silent Spring*. First, it argues that Carson chose to focus exclusively on the harmful effects of DDT, while disregarding its positive benefits, notably the control of the mosquito-borne disease malaria. Second, it points out that Carson ignored bird population data from the Audubon Society* that showed that many species of birds were actually increasing in population, rather than declining. Third, it claims that her data on the cancer* epidemic ignored significant statistical factors, such as an aging population and cancer cases caused by use of tobacco.⁷

Other scientific work continues to draw upon Carson's work, however. The US technology and culture scholar Edmund Russell's* *War and Nature* describes the relationship between chemical warfare and domestic pesticide use and draws directly on Carson's work. According to the British *Observer* newspaper's science editor, "Carson's warnings are still highly relevant, both in terms of the specific threat posed by DDT and its sister chemicals and to the general ecological dangers facing humanity." He goes on to cite several ecological examples of the continued presence of chemical pesticides in various forms of wildlife.[8]

Carson alerted the public to growing concerns within the scientific community and assigned them the moral obligation to question, learn, and take action. The public response following the book's publication was strong. Thousands of private citizens wrote to their local congressional representatives or senators requesting information and demanding that action be taken. Dozens of environmental advocacy groups were established in the years following *Silent Spring*'s publication. In 1970, fueled in part by Carson's work, the United States established its Environmental Protection Agency (EPA), a federal agency whose administrator is appointed directly by the president.[9] In 1980, President Jimmy Carter posthumously awarded Rachel Carson the Presidential Medal of Freedom, the highest civilian honor in the United States, for her work in bringing environmental concerns to public awareness.[10]

---

1. President's Science Advisory Committee (PSAC), *The Use of Pesticides*, May 15, 1963.
2. Toxic Substance Control Act, 15 US Code Chapter 53, 1976.
3. Property and Environment Research Center, "*Silent Spring* at 50: Reexamining Rachel Carson's Classic," accessed December 14, 2015, http://www.perc.org/blog/silent-spring-50-reexamining-rachel-carsons-classic.

4. Property and Environment Research Center, "*Silent Spring* at 50."
5. Eliza Griswold, "How 'Silent Spring' Ignited the Environmental Movement," *New York Times*, September 21, 2012.
6. Bryan Walsh, "How *Silent Spring* Became the First Shot in the War on the Environment," *Time*, September 25, 2012.
7. Roger Meiners, Pierre Desrochers, and Andrew Morriss, eds, *Silent Spring at 50: The False Crises of Rachel Carson* (Washington, DC: Cato Institute, 2012).
8. Robin McKie, "Rachel Carson and the Legacy of *Silent Spring*," *Guardian*, May 26, 2012.
9. Epa.gov, accessed December 5, 2015.
10. Jimmy Carter, "Presidential Medal of Freedom Remarks at the Presentation Ceremony," June 9, 1980. Online by Gerhard Peters and John T. Woolley, *The American Presidency Project*, accessed December 5, 2015, http://www. presidency.ucsb.edu/ws/?pid=45389.

MODULE 11
IMPACT AND INFLUENCE TODAY

KEY POINTS
- More than 50 years after its publication, *Silent Spring* remains a significant but controversial piece of environmental* literature.
- Carson's perspective on man's interaction with nature is still heavily debated within scientific and policy circles.
- It is still a challenge for environmentalist thinkers to discern objective truth when faced with conflicting and incomplete data.

Position

Rachel Carson's *Silent Spring* is still an important work for all those who are passionate about environmental issues and policies. The validity of Carson's specific claims regarding DDT* and other chemical pesticides and their long-term consequences continues to be heavily debated; currently Carson is generally regarded as a hero, though some people do find fault with her scientific and doom-and-gloom claims. Much of the critically hostile book *Silent Spring at 50* seeks to prove, with scientific data that compete with Carson's, that her claims were false and unsubstantiated, that she ignored any evidence that weakened her argument, and that she unnecessarily frightened the American public.[1]

While DDT has been banned in the United States since the early 1970s, the dangers of chemical pesticides and, more broadly, chemicals released into the environment are still a concern. Scientists, medical doctors, policymakers, and the general public continue to analyze the issue and form their own opinions on everything from BPA-free* bottles (plastic bottles made without the chemical Bisphenol A, implicated in birth defects) to how much soy is appropriate in one's diet and how many birds are killed by windmills. The underlying theme in all of these discussions is that man's impact on nature has uncertain, and sometimes dire, consequences. Furthermore, it is open to question as to which individuals and institutions have the authority and right to determine regulations and set the levels of risk and exposure. Perhaps Carson's most important legacy is introducing to the public the idea that they need to think for themselves.

> "Carson's book was controversial before it even was a book."
> ——William Souder,* "Rachel Carson Didn't Kill Millions of Africans"

Interaction

In his introduction to the 1994 edition of *Silent Spring,* the then vice president of the United States Al Gore* said, "Writing about *Silent Spring* is a humbling experience for an elected official, because Rachel Carson's book provides undeniable proof that the power of an idea can be far greater than the power of politicians." He goes on to give Carson credit for engaging him in environmental concerns. Al Gore and the Intergovernmental Panel on Climate Change (IPCC)* won the Nobel Peace Prize in 2007, in part for their mainstream blockbuster hit documentary, *An Inconvenient Truth.** The film brought the issue of climate change into popular culture and showed how industrial actions and development have shifted the earth's climate patterns since the Industrial Revolution.*

Carson's supporters and critics continue to debate the impact *Silent Spring* had on the banning of DDT. Ironically it seems that it is her critics rather than her supporters who attribute this victory to her. Carson's critics blame her for "millions of African deaths" caused by malaria because her book, they claim, prevented DDT from being used to kill mosquito populations in Africa.² Her supporters, on the contrary, claim DDT was already beginning to decline in popularity by the time *Silent Spring* was published, at least in the United States, because massive spraying initiatives had led to the development of a DDT-resistant strain of mosquito. In 2006, the World Health Organization,* the division of the United Nations concerned with public health, began to revisit its DDT-spraying initiatives to combat malaria in Africa, where, Carson's biographer William Souder* points out, DDT has never been banned.

Carson's ideas are still current in the ongoing debates on climate change. Environmental scholars such as Bill McKibben* and Amory Lovins* of the United States insist we must take action to reduce atmospheric carbon and our dependence on fossil fuels. Critics of geoengineering* (the field of engineering theory that focuses on planet-scale technological solutions to mitigate the impacts of climate change and sea-level rise) cite the precautionary principle*—that it is better to limit interference with the planet as we can never be sure of the ultimate effects. Meanwhile, consumers and citizens continue to demand greater information and transparency in the products they buy and the companies they support.

The Continuing Debate

Carson's biggest opponents are the usual opponents of environmental regulation—typically those focused on private property rights, capitalist* economic markets, big business, and conservative politics (capitalism is the social and economic model dominant in the West and increasingly throughout the world, in which trade and industry are conducted for private profit). Unsurprisingly, chemical manufacturers continue to refute her claims, since *Silent Spring* posed a direct threat to their livelihood. "Carson's 'you can't be too safe' standard is seen today in the 'precautionary principle' that helps to retard the adoption of superior technology that would benefit people and the environment," said Roger Meiners, distinguished professor of economics and law at the University of Texas at Arlington, senior fellow at the Property and Environment Research Center,* and one of the three main editors of *Silent Spring at 50*.[3] He believes her simplified view of risk appears to have affected the drafting of the US federal government's Clean Air Act* and Clean Water Act* that set "impossible standards in some areas not remotely related to human health or technical feasibility."

Debates about environmental regulation, economic growth, and the precautionary principle continue to dominate every global conference on environmental matters, including the conferences on climate run under the auspices of the United Nations such as the

Earth Summits in Rio of 1992 and 2012, and the United Nations Conference on Sustainable Development (Rio + 20). The debate will continue because there is no clear answer and the facts and circumstances continue to change. Undeniably, Carson was and remains a significant part of this conversation.

---

1. Roger Meiners, Pierre Desrochers, and Andrew Morriss, eds, *Silent Spring at 50: The False Crises of Rachel Carson* (Washington, DC: Cato Institute, 2012).
2. William Souder, *On a Farther Shore: The Life and Legacy of Rachel Carson* (New York: Broadway Books, 2012), 332–5.
3. https://www.masterresource.org/silent-spring-at-50/silent-spring-at-50/.

MODULE 12
WHERE NEXT?

KEY POINTS
- *Silent Spring* shows the importance of analyzing humanity's influence on the environment.*
- Carson's quest to alert the public to industrial practices and the loopholes in environmental regulation remains relevant today.
- *Silent Spring* acts as a call to action for anyone concerned about the long-term sustainability of the planet for future generations, regardless of whether one accepts or rejects Carson's particular claims.

Potential

Rachel Carson's *Silent Spring* is still a powerful force in bringing to light concerns over pesticide use and the environment generally. Her approach remains relevant to the political and scientific challenges presented by global climate change,* the release of gasses that trap solar energy, and energy security (a nation's assured access to energy or fuel). World leaders debate extreme ideas about geoengineering* while conservationists remind us about the precautionary principle.*

Silent Spring remains alive in these debates and scholars continue to discuss Rachel Carson's impact on fields as diverse as environmental regulation, government leadership, feminism,* environmental conservation,* and morality (ethical behavior).

Silent Spring continues to be a highly esteemed piece of nonfiction literature. It was named one of the "25 Greatest Scientific

Books of All Time" by *Discover* magazine,[1] and the *Guardian*, a UK newspaper, listed it as part of its "Fifty Books to Change the World."[2] It is included in the "100 Best Nonfiction Books of the Twentieth Century," compiled by the *National Review*, and it has a place on *Time* magazine's "All Time Greatest Nonfiction Books" listing. The number of environmental science,* study, and policy programs in higher education continues to increase, and most consider *Silent Spring* required reading. Regardless of one's particular opinion about Carson's claims, *Silent Spring* is necessary reading for anyone seeking to understand modern environmental literature, policy, and culture.

Perhaps more than any specific argument offered by Carson, her most important idea is that we, as a species, can never fully understand the impact we have on the environment, and to think otherwise is naïve. The mysteries of the oceans and the nuanced, secret wonders of the natural world Carson revered can never be fully modeled, regardless of how advanced the algorithm or computing device. Some aspects of nature will always remain unknown, and Carson encourages us to embrace this wonder with respect and humility.

> "And there the two sides sit 50 years later. On one side of the environmental debate are the perceived softhearted scientists and those who would preserve the natural order; on the other are the hard pragmatists [realists] of industry and their friends in high places, the massed might of the establishment. Substitute climate change for pesticides,* and the argument plays out the same now as it did a half-century ago."
> ——William Souder, "Rachel Carson Didn't Kill Millions of Africans"

Future Directions

Carson's work continues to inspire heated debate and ongoing scholarly research. The 50th anniversary of *Silent Spring's* publication prompted a deluge of attention and analysis. Carson continues to challenge critics and inspire the next generation of environmental leaders. Much of the future analysis and continuation of her work will be supported by institutions dedicated to preserving her memory and, conversely, by those that challenge it.

Several nonprofit institutions have been established in her memory. The Rachel Carson Institute at Carson's alma mater, Chatham College, "continues the legacy of Chatham's most famous alumna, Rachel Carson, Class of '29, author, scientist and credited with helping form the modern environmental movement."*[3] The Institute hosts a number of programs in her memory that are designed to help continue the values, goals, and directions of her work and support future generations of environmental scholars and leaders. Similar institutions include the Silent Spring Institute, a community of "researchers dedicated to science that serves the public interest,"[4] funded by a number of notable government institutes and private foundations. Ludwig Maximillian University in Munich hosts the Rachel Carson Center for Environment and Society, there is a national wildlife refuge in coastal Maine that bears her name, and her childhood home is maintained as a museum. In short, Carson lives on in American and international cultural memory.

In contrast, there are several activist groups that continue to critique her work and what she stood for. The website *rachelwaswrong.com* continues to claim that Carson was directly responsible for the deaths of millions of Africans from malaria, due to *Silent Spring*'s call for the regulation of DDT. It states, "This website addresses the dangers associated with anti-technology views, as embodied in Rachel Carson's *Silent Spring*. Such views pervade much of modern-day environmental literature, and have actually become part of the world's conventional wisdom."[5]

Summary

Today, Rachel Carson's *Silent Spring* forms a fundamental part of environmental literature and plays a fundamental role in the larger environmental movement. For those who embrace her ideas and goals, she will continue to be a source of inspiration and leadership, reminding us to embrace the beauty, wonder, and fragility of the natural world and to treat all living creatures with respect. For her critics, she will remain someone whose ideas must be refuted and who is a testament to the idea that society can succumb to the power of graceful words. For everyone, she continues to remind us of the need to question the status quo, to think for oneself, and to speak up and out about the issues that matter most.

Silent Spring presents the reader with a chilling warning of the dangers of a highly industrialized and growth-focused society, painting a picture of a natural world that could easily be destroyed or lost if humanity continues to act without understanding the consequences of those actions. *Silent Spring* teaches the reader to embrace wonder and science at the same time, rather than see them as opposing forces, and to understand human beings and modern society as part of, not enemies of, the natural world. It teaches the reader to ask questions and understand the answers: "If, having endured much, we have at last asserted our right to know, and if, knowing, we have concluded that we are being asked to take senseless and frightening risks, then we should no longer accept the counsel of those who tell us that we must fill our world with poisonous chemicals; we should look about and see what other course is open to us."[6]

---

1. "25 Greatest Scientific Books of All Time," *Discover*, December 8, 2006.
2. "Fifty Books to Change the World," *Guardian*, January 27, 2010.
3. Rachel Carson Institute, Chatham University, accessed December 12, 2015, http://www.chatham.edu/centers/rachelcarson/.
4. Silent Spring Institute, accessed December 12, 2015, http://www.silentspring.org/.
5. Rachel Was Wrong, accessed December 12, 2015, www.rachelwaswrong.org.
6. Rachel Carson, *Silent Spring* (New York: Houghton Mifflin Harcourt, 2002), 277–8.

GLOSSARY OF TERMS

1. ***An Inconvenient Truth***: a 2006 documentary film starring the then United States vice president Al Gore that reviewed the science behind climate change and its potential effects. It was award the 2007 Academy Award for best documentary film.

2. **Atom bomb:** a highly destructive nuclear weapon, developed in the 1940s; it has been used in warfare twice, in the destruction of the Japanese cities of Hiroshima and Nagasaki in August 1945.

3. **Audubon Society:** an influential conservation advocacy group, founded in the United States in 1905.

4. **Bioaccumulation:** the process of absorbing and storing any type of chemical or compound at a rate faster than it is dissipated. It can also refer to the compounding impacts of a substance as it moves up the food chain when organisms consume contaminated prey.

5. **Biocide:** any type of chemical compound capable of killing living creatures or plants. Specific to European environmental legislation, it is defined as a "chemical substance or microorganism intended to destroy, deter, render harmless, or exert a controlling effect on any harmful organism by chemical or biological means."

6. **Biological pest control (biological control of insects):** the practice of managing landscape and agricultural pests through the use of natural biological controls such as parasites, predatory insects or pathogens, which reduces the need for chemical pesticides.

7. **Biology:** the scientific study of living organisms.

8. **Biotechnology:** the use of organisms, and biological matter more generally, in the construction of tools useful to human purposes.

9. **BPA (Bisphenol A):** a chemical compound found in plastics and epoxy resins commonly used to coat the inside of food containers such as metal cans and plastic food containers and water bottles. Some research suggests that BPA can leach into food or liquid stored in BPA-lined containers, and that consumption of BPA can have adverse effects on children and unborn fetuses.

10. **Cancer:** a disease, or group of diseases, characterized by abnormal cell growth.

11. **Capitalism:** the social and economic model dominant in the West and increasingly throughout the world, in which trade and industry are conducted for private profit.

12. **Carcinogen:** a substance known to cause, or have the potential to cause, cancer.

13. **Cato Institute:** a right-wing think tank, based in Washington, DC, advocating libertarian politics.

14. **CBS:** a broadcaster in the United States; the initials stand for "Colombia Broadcast System."

15. **Clean Air Act (1970):** a United States law that seeks to improve, strengthen, and accelerate programs for the prevention and abatement of air pollution.

16. **Clean Water Act (1972):** a United States law that comprehensively addresses water quality and pollution.

17. **Communist:** someone who subscribes to the political ideology of communism, which relies on the state ownership of the means of production, the collectivization of labor, and the abolition of social class.

18. **Competitive Enterprise Institute (CEI):** a think tank founded in the United States in 1984 in Washington, DC. It advocates for libertarian economic policies, through the restriction of government regulation.

19. **Conservation movement (in America, 1890–1920):** a political and social movement with a focus on conservation, environmental protection, and a celebration of national landscape resources such as National Parks and National Forests.

20. **Crop rotation:** the agricultural method of rotating the type of crop planted in a given field or garden over a given period of time; benefits include increased soil fertility and pest control.

21. **DDT (dichloro-diphenyl-trichloroethane):** an organochlorine compound with insecticidal properties that was mainly used to control mosquito-borne malaria. Although it is effective in destroying certain living things that are harmful to animals and plants, it can also be extremely dangerous to humans and the environment.

22. **Defoliation:** stripping a tree of its leaves or otherwise causing a tree to lose the majority of its leaf cover, typically through pesticide application or insect infestation.

23. **Earth Day:** begun in 1970 and held every year on April 22, Earth Day is held to draw attention to the importance of environmental concerns and protection.

24. **Earth stewardship:** a concept promoted by the Ecological Society of America that examines both socioecological change and ecosystem resilience at all scales to enhance human well-being.

25. **Ecofeminism:** a loosely defined political movement of conservation and environmental stewardship promoted by women (particularly American women) beginning in the 1960s and 1970s, that typically combines conservation advocacy and mainstream feminist thought with earth spirituality. Key figures include Mary Daly, Susan Griffin, Ellen Willis, and Rachel Carson.

26. **Ecology:** a branch of biology that studies the way groups of living things interact with one another and with their environment.

27. **Ecosystem:** a community of living organisms in a given location.

28. **Endocrine disruptors/disruption:** a category of chemicals and substances that can interrupt the natural processes of the body's endocrine hormones.

29. **Environment:** the setting or condition in which a person, plant or animal lives or operates.

30. **Environmental conservation:** the practice of using environmental resources judiciously and ensuring continued use of environmental and natural resources for future generations while preserving the health and resilience of ecosystems.

31. **Environmental movement:** the (modern) environmental movement in the United States began in the 1960s and is a political, scientific, and social movement focused on bringing attention to environmental concerns and developing strategies to improve environmental and ecological circumstances.

32. **Environmental Protection Agency (EPA):** an agency in the executive branch of the United States government focused on environmental health, quality, and regulation, with branches at the federal, state, and local level.

33. **Environmental science:** scientific study of the environment, commonly conducted by drawing on disciplines such as biology, physics, and ecology.
34. **Feminism:** the political and cultural movement associated with the struggle for equality between the sexes.
35. **Genetics:** the science dealing with genes, the biological material allowing the transmission of characteristics from generation to generation.
36. **Geoengineering:** also known as climate engineering, this refers to the modification of the earth's climate systems via artificial means in an attempt to control atmospheric carbon and reduce the predicted impacts of climate change.
37. **Global climate change:** long-term change in patterns governing the earth's weather and temperature.
38. **Gypsy moth** (*Lymantria dispar dispar*)**:** a moth introduced into the United States from Europe. The larvae feed on the foliage of shade and other types of trees.
39. **Industrial Revolution:** a period of intense technological and industrial growth via new manufacturing processes beginning in Britain in the 1760s with the mechanization of textile manufacturing. Several climate models indicate that this was the period in which atmospheric carbon began increasing at non-natural rates.
40. **Insect:** consisting of more than a million species, this is the most diverse group of animals, representing more than half of all known living organisms.
41. **Insect-borne illness:** a disease carried and transmitted to humans by insects.
42. **IPCC:** the Intergovernmental Panel on Climate Change, a branch of the United Nations, was formed in 1988, with its primary goal being to "stabilize greenhouse gas concentrations in the atmosphere at a level that would prevent dangerous anthropogenic [human-induced] interference with the climate system."
43. **Land ethic:** a term in Aldo Leopold's book *Sand County Almanac* that calls for a new, more respectful, and more thoughtful relationship between man and the land he inhabits.

| Glossary of Terms

44. **Leukemia:** a group of cancers that begin in the bone marrow.

45. **Mammal (mammalia):** a type of vertebrate characterized by having hair, three distinct middle ear bones, mammary glands to nourish their young with milk, and a neocortex (the part of the cerebral cortex involved in sight and hearing). Over 4,000 distinct mammalian species have been identified.

46. **Monoculture:** the standard agricultural practice of growing a single species over a large area. It maximizes efficiency but can make crops more vulnerable to insects or diseases.

47. **Monsanto:** a multinational agricultural chemical and biotechnology company founded in the United States in 1901.

48. **National Agricultural Chemicals Association:** an organization founded in the United States and concerned with the usage, production, and environmental effects of agricultural chemicals such as fertilizers and pesticides.

49. **National Cancer Institute (NCI):** a government agency, forming one of the 11 agencies of the US government's Department of Health and Human Services. It was founded to further research into the causes of and treatments for cancer, and for the supporters of those suffering from and affected by cancer.

50. **National Environmental Protection Act (1970):** a United States law that established the White House Council on Environmental Quality and promoted the idea that environmental impacts should be given equal consideration when making policy decisions.

51. **National Institutes of Health (NIH):** a federal agency of the US Department of Health and Human Services (USDHHS) and the primary medical research agency within the federal government.

52. **Natural pesticides:** a nonchemical or homeopathic substance used for pest control, many times in conjunction with biological pest control strategies.

53. **Pesticide:** any type of substance used to control or kill pests, in an effort either to protect agricultural crops or to maintain safety and cleanliness for human inhabitation.

54. **Precautionary principle:** an approach to risk assessment that emphasizes

caution in the face of uncertainty or that prevents action without adequate knowledge of the consequences.

55. **Property and Environment Research Center:** an organization founded in the United States in 1982 seeking market solutions to environmental problems.
56. **Radiation:** the emission or transmission of energy through particle movement, typically electromagnetic waves.
57. **Resistance (to pesticides):** the phenomenon of insects becoming less impacted by a given pesticide or compound. In a given insect population, some specimens will naturally be resistant to a pesticide. The release of a pesticide will first kill off the weakest individuals, thus allowing those resistant to the compound to breed with reduced competition, resulting in a growing population of pesticide-resistant individuals.
58. **Smithsonian Institution:** the world's largest group of museums and research centers, administered by the United States federal government and established in 1846.
59. **Soviet Union:** a federation of communist states that existed between 1922 and 1991, centered primarily on Russia and its neighbors in Eastern Europe and the northern half of Asia.
60. **Synthetic pesticides:** man-made chemicals (or mixtures of chemicals) developed to prevent or destroy certain organisms (insects, weeds, or fungi) deemed "pests."
61. **Toxic Substance Control Act (1976):** a United States law administered by the Environmental Protection Agency that, in the words of the act, "regulates commerce and protects human health and the environment by requiring testing and necessary use restrictions on certain chemical substances, and for other purposes."
62. **US Bureau of Fisheries (now the US Fish and Wildlife Service):** the main conservation agency within the Department of the Interior.
63. **US Fish and Wildlife Service:** founded as the US Bureau of Fisheries, the US Fish and Wildlife Service is the main conservation agency within the United

States Department of the Interior.

64. **World Health Organization:** the division of the United Nations responsible for global public health.

65. **World War II (1939–45):** a global war between two major international alliances—the Axis (Germany, Italy, Japan) and the Allies (led by the Soviet Union, the United States, and Great Britain)—and considered the deadliest conflict in human history.

66. **Zoology:** the study of animal life.

PEOPLE MENTIONED IN THE TEXT

1. **Erin Brockovich (b. 1960)** was an American legal clerk who helped bring about one of the largest lawsuits against a utility company—Pacific Gas and Electric in 1993, for its use of hexavalent chromium. Her story was popularized by the movie *Erin Brockovich*, released in 2000 with Julia Roberts in the lead role.
2. **Ralph Waldo Emerson (1803–82)** was an American poet and author.
3. **Howard Frech** was an American environmentalist artist. He worked for the *Baltimore Sun*, where Carson met him.
4. **Albert "Al" Gore (b. 1948)** was awarded the Nobel Peace Prize in 2007 for his work in climate-change activism. He served as the 45th vice president of the United States from 1993 to 2001 under President Bill Clinton.
5. **Eliza Griswold (b. 1973)** is a widely published American journalist and poet, and the author of the *New York Times* article "How 'Silent Spring' Ignited the Environmental Movement" (2012).
6. **Wilhelm Carl Hueper (1894–1978)** was the first director of the Environmental Cancer Section of the National Cancer Institute from 1938 to 1964. He became a role model for environmental scientist and author Rachel Carson, and collaborated with her on research for *Silent Spring*.
7. **John F. Kennedy (1917–63)** was a member of the Democratic Party and 35th president of the United States. He served in office from 1961 to 1963.
8. **Linda Lear (b. 1940)** is an American writer who published *Rachel Carson: Witness for Nature* in 1997.
9. **Aldo Leopold (1887–1948)** was an author and environmentalist from the United States. He is noted for his *Sand County Almanac*, a work calling for a more respectful and thoughtful relationship between human beings and the land we inhabit.
10. **Amory Lovins (b. 1947)** is an American physicist and environmental activist who is currently chief scientist at the Rocky Mountain Institute, an environmental research and consulting organization. He featured in *Time* magazine's list of the most influential people of 2009.

11. **Mark Hamilton Lytle** is professor of history and environmental studies at Bard College in New York state. Among his works is *The Gentle Subversive* (1988), an appraisal of Rachel Carson's legacy.

12. **Bill McKibben (b. 1960)** is an American author and environmentalist who cofounded 350.org, an anti-carbon organization focused on reducing the concentration of atmospheric carbon to below 350 parts per million.

13. **Margaret Mead (1901–78)** was an American anthropologist and writer focused on environmental issues in the 1960s and 1970s.

14. **Joni Mitchell (b. 1943)** is a Canadian singer-songwriter. Beginning as a folk singer in the 1960s, she moved on to more experimental work in the 1970s, drawing on the traditions of forms such as jazz and African music.

15. **John Muir (1838–1914)** was a Scottish American naturalist whose activism was instrumental in the protection of areas of wilderness such as Yosemite Valley and the Sequoia National Park in the United States. He was an early (and notably influential) author in the tradition of environmental literature.

16. **Paul "Pauly" Müller (1899–1965)** was a Swiss chemist noted as the inventor of the pesticide DDT in 1939, for which he received the 1948 Nobel Prize in Physiology or Medicine.

17. **Theodore "Teddy" Roosevelt (1858–1919)** was a member of the Republican Party and 26th president of the United States, who served in office from 1901 to 1909.

18. **Edmund Russell** is an assistant professor in the Division of Technology, Culture, and Communication in the School of Engineering and Applied Science at the University of Virginia.

19. **Charles Schultz (1922–2000)** was an award-winning American cartoonist who was best known for the *Peanuts* characters.

20. **William Souder** is an American biographer; he is the author of *On a Farther Shore: The Life and Legacy of Rachel Carson* (2013).

21. **Harriet Beecher Stowe (1811–96)** was an American author and activist in the struggle against the institution of slavery. She is noted for her novel *Uncle*

Tom's Cabin (1852), which brought the plight of African American slaves to the attention of millions around the world.

22. **Henry David Thoreau (1817–62)** was an American political theorist, author, activist, and poet, noted for coining the phrase "civil disobedience" and for his book *Walden* (1854)—an early text in the tradition of modern environmental literature.

WORKS CITED

1. "25 Greatest Scientific Books of All Time." *Discover*, December 8, 2006.

2. Belsie, Laurent. "Gypsy Moths Return to Northeast: Worst Outbreak in a Decade Descends on Northeast; Entomologists Do Not Know How to Stop It." *Christian Science Monitor*, July 2, 1990.

3. Brinkley, Douglas. *The Wilderness Warrior: Theodore Roosevelt and the Crusade for America*. New York: Harper Perennial, 2010.

4. Carson, Rachel. *Under the Sea Wind*. New York: Simon & Schuster, 1941.

5. *The Sea Around Us*. Oxford: Oxford University Press, 1951.

6. *The Edge of the Sea*. Boston: Houghton Mifflin, 1955.

7. *Silent Spring*. New York: Houghton Mifflin Harcourt, 1994, 2002.

8. Davis, Devra. *The Secret History of the War on Cancer*. New York: Basic Books, 2007.

9. "Fifty Books to Change the World." *Guardian*, January 27, 2010.

10. Gillam, Scott. *Rachel Carson: Pioneer of Environmentalism*. Edina, MN: ABDO, 2011.

11. Griswold, Eliza. "How 'Silent Spring' Ignited the Environmental Movement." *New York Times*, September 21, 2012.

12. Hueper, Wilhelm C. "Occupational Tumors and Allied Diseases." *Occupational Tumors and Allied Diseases* (1942).

13. Hynes, H. Patricia. "Unfinished Business: *Silent Spring* on the 30th Anniversary of Rachel Carson's Indictment of DDT, Pesticides Still Threaten Human Life." *Los Angeles Times*, September 10, 1992.

14. Keller, Julia. "Works Such as *Uncle Tom's Cabin* and *Silent Spring* Had Such a Profound Political Effect that They Became ... Art that Changed the World." *Chicago Tribune*, June 27, 1999.

15. Kroll, Gary. "The 'Silent Springs' of Rachel Carson: Mass Media and the Origins of Modern Environmentalism." *Public Understanding of Science* 10, no. 4 (2001): 403–20.

16. Lear, Linda. *Rachel Carson: Witness for Nature*. New York: Macmillan, 1998.
17. RachelCarson.org. Accessed December 12, 2015. http://www.rachelcarson.org/SeaAroundUs.aspx.
18. Leopold, Aldo. *Sand County Almanac: And Sketches Here and There*. Oxford: Oxford University Press, 1949.
19. Lytle, Mark Hamilton. *The Gentle Subversive: Rachel Carson, Silent Spring, and the Rise of the Environmental Movement*. New York and Oxford: Oxford University Press, 2007.
20. McKie, Robin. "Rachel Carson and the Legacy of *Silent Spring*." *Guardian*, May 26, 2012.
21. Meiners, Roger, Pierre Desrochers, and Andrew Morriss, eds. *Silent Spring at 50: The False Crises of Rachel Carson*. Washington, DC: Cato Institute, 2012.
22. Milne, Lorus, and Margery Milne. "There's Poison All Around Us Now." *New York Times*, September 23, 1962.
23. Monsanto Corporation. "The Desolate Year." *Monsanto Magazine* (October 1962).
24. Moyers, Bill. *PBS Journal*, September 21, 2007. Accessed 21 February 2016. http://www.pbs.org/moyers/journal/09212007/watch.html.
25. Norwood, Vera L. "The Nature of Knowing: Rachel Carson and the American Environment." *Signs* (1987): 740–60.
26. Oelschlaeger, Max. "Emerson, Thoreau, and the Hudson River School." *Nature Transformed*, National Humanities Center. Accessed December 30, 2015. http://nationalhumanitiescenter.org/tserve/nattrans/ntwilderness/essays/preserva.htm.
27. President's Science Advisory Committee (PSAC). *The Use of Pesticides*. May 15, 1963.
28. Property and Environment Research Center. "*Silent Spring* at 50: Reexamining Rachel Carson's Classic." Accessed December 14, 2015. http://www.perc.org/blog/silent-spring–50–reexamining-rachel-carsons-classic.
29. Quaratiello, Arlene Rodda. *Rachel Carson: A Biography*. Westport, CT:

Greenwood Press, 2010.
30. Rothman, Joshua. "Rachel Carson's Natural Histories." *New Yorker*, September 27, 2012.
31. Rubin, Charles T. *The Green Crusade*. Lanham, MD: Rowman & Littlefield, 1994.
32. Russell, Edmund. *War and Nature: Fighting Humans and Insects with Chemicals from World War I to Silent Spring*. New York: Cambridge University Press, 2001.
33. Schuldenfrei, Eric. *The Films of Charles and Ray Eames: A Universal Sense of Expectation*. New York: Routledge, 2015.
34. Souder, William. *On a Farther Shore: The Life and Legacy of Rachel Carson*. New York: Broadway Books, 2012.
35. "Rachel Carson Didn't Kill Millions of Africans." *Slate*, September 4, 2012.
36. Stowe, Harriet Beecher. *Uncle Tom's Cabin*. Leipzig: Tauchnitz, 1852.
37. Walsh, Bryan. "How *Silent Spring* Became the First Shot in the War over the Environment." *Time*, September 25, 2012.
38. Willer, Robb. "Is the Environment a Moral Cause?" *New York Times*, February 27, 2015.
39. Zubrin, Robert. "The Truth about DDT and *Silent Spring*." *The New Atlantis.com*, September 27, 2012. Accessed March 4, 2016. http://www.thenewatlantis.com/publications/the-truth-about-ddt-and-silent-spring.

原书作者简介

蕾切尔·卡森生于 1907 年，她在美国宾夕法尼亚州斯普林代尔区的家庭农场度过了童年时光。她的乡村成长经历奠定了她与自然界深层联系的基础，她后来之所以成为最早的环保作家和环保活动家之一，显然与此相关。作为学者，卡森先后在宾夕法尼亚女子学院、伍兹霍尔学院和约翰·霍普金斯大学学习生物学和动物学，后在美国政府部门科学岗位任职 15 年。此后，卡森将余生投入写作并创作出四部环境主题的畅销书。1964 年她因乳腺癌辞世，年仅 56 岁。

本书作者简介

妮姬·斯普林格曾先后就读于麻省理工学院、哈佛大学和耶鲁大学。目前在耶鲁大学攻读环境管理学博士学位。

世界名著中的批判性思维

《世界思想宝库钥匙丛书》致力于深入浅出地阐释全世界著名思想家的观点，不论是谁、在何处都能了解到，从而推进批判性思维发展。

《世界思想宝库钥匙丛书》与世界顶尖大学的一流学者合作，为一系列学科中最有影响的著作推出新的分析文本，介绍其观点和影响。在这一不断扩展的系列中，每种选入的著作都代表了历经时间考验的思想典范。通过为这些著作提供必要背景、揭示原作者的学术渊源以及说明这些著作所产生的影响，本系列图书希望让读者以新视角看待这些划时代的经典之作。读者应学会思考、运用并挑战这些著作中的观点，而不是简单接受它们。

ABOUT THE AUTHOR OF THE ORIGINAL WORK

Rachel Carson was born in 1907 and spent her childhood on the family farm in Springdale, Pennsylvania, in the United States. Her rural upbringing laid the foundations for a deep connection with the natural world, and this would emerge later when Carson became one of the earliest environmentalist writers and campaigners. As an academic, Carson studied biology and later zoology at the Pennsylvania College for Women, Woods Hole Institute, and Johns Hopkins University. She then held scientific positions with the US government for 15 years. Carson devoted her later years to writing and produced four best sellers on environmental themes. She died in 1964 of breast cancer, aged just 56.

ABOUT THE AUTHOR OF THE ANALYSIS

Nikki Springer has studied at MIT, Harvard and Yale. She is currently researching her PhD in environmental management at Yale.

ABOUT MACAT
GREAT WORKS FOR CRITICAL THINKING

Macat is focused on making the ideas of the world's great thinkers accessible and comprehensible to everybody, everywhere, in ways that promote the development of enhanced critical thinking skills.

It works with leading academics from the world's top universities to produce new analyses that focus on the ideas and the impact of the most influential works ever written across a wide variety of academic disciplines. Each of the works that sit at the heart of its growing library is an enduring example of great thinking. But by setting them in context — and looking at the influences that shaped their authors, as well as the responses they provoked — Macat encourages readers to look at these classics and game-changers with fresh eyes. Readers learn to think, engage and challenge their ideas, rather than simply accepting them.

批判性思维和《寂静的春天》

首要批判性思维技能：理性化思维

次要批判性思维技能：创造性思维

蕾切尔·卡森的《寂静的春天》（1962）是少有的可以称得上具有划时代意义的一部著作。它对美国农用杀虫剂的使用发起了论证严密的抨击，将环保意识推到了现代政治和决策的显著位置，从而形成了当今我们所了解的环境监管局面。

该书也是展现严密论证之力量的典范。这一论证是通过组织严密、论证严谨的观点建立起来的。这些观点极具说服力，且设计得无可辩驳。事实上，也必须如此。因为，该书出版后，化工业利用一切资源试图诋毁《寂静的春天》和卡森本人——但徒劳无益。

该书的中心论断是：战后农业和化工业的进步促使人们无差别地使用杀虫剂。但这种无差别的使用对植物、动物乃至整个环境都造成了巨大伤害，其毁灭性的影响远远超过其对作物的保护作用。当时，这个论断与政府政策及正统的科学观念完全相左。很多证实卡森观点的研究都受到了充满敌意的商业利益集团的蓄意压制。然而，卡森在《寂静的春天》里收集、组织并阐述证据，以清晰确切的方式论证其观点。

虽然环境之战还在继续，但当今很少有人会否认她论证的力量和说服力。

CRITICAL THINKING AND *SILENT SPRING*

- Primary critical thinking skill: REASONING
- Secondary critical thinking skill: CREATIVE THINKING

Rachel Carson's 1962 *Silent Spring* is one of the few books that can claim to be epoch-making. Its closely reasoned attack on the use of pesticides in American agriculture helped thrust environmental consciousness to the fore of modern politics and policy, creating the regulatory landscape we know today.

The book is also a monument to the power of closely reasoned argument built from well organised and carefully evidenced points that are not merely persuasive, but designed to be irrefutable. Indeed, it had to be: upon its publication, the chemical industry utilised all its resources to attempt to discredit both *Silent Spring* and Carson herself — to no avail.

The central argument of the book is that the indiscriminate use of pesticides encouraged by post-war advances in agriculture and chemistry was deeply harmful to plants, animals and the whole environment, with devastating effects that went far beyond protecting crops. At the time, the argument directly contradicted government policy and scientific orthodoxy, and many studies that corroborated Carson's views were deliberately suppressed by hostile business interests. Carson, however, gathered, organised and set out the evidence in *Silent Spring* in a way that proved her contentions without a doubt.

While environmental battles still rage, few now deny the strength and persuasiveness of her reasoning.

《世界思想宝库钥匙丛书》简介

《世界思想宝库钥匙丛书》致力于为一系列在各领域产生重大影响的人文社科类经典著作提供独特的学术探讨。每一本读物都不仅仅是原经典著作的内容摘要,而是介绍并深入研究原经典著作的学术渊源、主要观点和历史影响。这一丛书的目的是提供一套学习资料,以促进读者掌握批判性思维,从而更全面、深刻地去理解重要思想。

每一本读物分为3个部分:学术渊源、学术思想和学术影响,每个部分下有4个小节。这些章节旨在从各个方面研究原经典著作及其反响。

由于独特的体例,每一本读物不但易于阅读,而且另有一项优点:所有读物的编排体例相同,读者在进行某个知识层面的调查或研究时可交叉参阅多本该丛书中的相关读物,从而开启跨领域研究的路径。

为了方便阅读,每本读物最后还列出了术语表和人名表(在书中则以星号 * 标记),此外还有参考文献。

《世界思想宝库钥匙丛书》与剑桥大学合作,理清了批判性思维的要点,即如何通过6种技能来进行有效思考。其中3种技能让我们能够理解问题,另3种技能让我们有能力解决问题。这6种技能合称为"批判性思维PACIER模式",它们是:

分析:了解如何建立一个观点;
评估:研究一个观点的优点和缺点;
阐释:对意义所产生的问题加以理解;
创造性思维:提出新的见解,发现新的联系;
解决问题:提出切实有效的解决办法;
理性化思维:创建有说服力的观点。

THE MACAT LIBRARY

The Macat Library is a series of unique academic explorations of seminal works in the humanities and social sciences — books and papers that have had a significant and widely recognised impact on their disciplines. It has been created to serve as much more than just a summary of what lies between the covers of a great book. It illuminates and explores the influences on, ideas of, and impact of that book. Our goal is to offer a learning resource that encourages critical thinking and fosters a better, deeper understanding of important ideas.

Each publication is divided into three Sections: Influences, Ideas, and Impact. Each Section has four Modules. These explore every important facet of the work, and the responses to it.

This Section-Module structure makes a Macat Library book easy to use, but it has another important feature. Because each Macat book is written to the same format, it is possible (and encouraged!) to cross-reference multiple Macat books along the same lines of inquiry or research. This allows the reader to open up interesting interdisciplinary pathways.

To further aid your reading, lists of glossary terms and people mentioned are included at the end of this book (these are indicated by an asterisk [*] throughout) — as well as a list of works cited.

Macat has worked with the University of Cambridge to identify the elements of critical thinking and understand the ways in which six different skills combine to enable effective thinking.

Three allow us to fully understand a problem; three more give us the tools to solve it. Together, these six skills make up the PACIER model of critical thinking. They are:

ANALYSIS — understanding how an argument is built
EVALUATION — exploring the strengths and weaknesses of an argument
INTERPRETATION — understanding issues of meaning
CREATIVE THINKING — coming up with new ideas and fresh connections
PROBLEM-SOLVING — producing strong solutions
REASONING — creating strong arguments

"《世界思想宝库钥匙丛书》提供了独一无二的跨学科学习和研究工具。它介绍那些革新了各自学科研究的经典著作,还邀请全世界一流专家和教育机构进行严谨的分析,为每位读者打开世界顶级教育的大门。"

—— 安德烈亚斯·施莱歇尔,
经济合作与发展组织教育与技能司司长

"《世界思想宝库钥匙丛书》直面大学教育的巨大挑战……他们组建了一支精干而活跃的学者队伍,来推出在研究广度上颇具新意的教学材料。"

—— 布罗尔斯教授、勋爵,剑桥大学前校长

"《世界思想宝库钥匙丛书》的愿景令人赞叹。它通过分析和阐释那些曾深刻影响人类思想以及社会、经济发展的经典文本,提供了新的学习方法。它推动批判性思维,这对于任何社会和经济体来说都是至关重要的。这就是未来的学习方法。"

—— 查尔斯·克拉克阁下,英国前教育大臣

"对于那些影响了各自领域的著作,《世界思想宝库钥匙丛书》能让人们立即了解到围绕那些著作展开的评论性言论,这让该系列图书成为在这些领域从事研究的师生们不可或缺的资源。"

—— 威廉·特朗佐教授,加利福尼亚大学圣地亚哥分校

"Macat offers an amazing first-of-its-kind tool for interdisciplinary learning and research. Its focus on works that transformed their disciplines and its rigorous approach, drawing on the world's leading experts and educational institutions, opens up a world-class education to anyone."

—— Andreas Schleicher, Director for Education and Skills, Organisation for Economic Co-operation and Development

"Macat is taking on some of the major challenges in university education... They have drawn together a strong team of active academics who are producing teaching materials that are novel in the breadth of their approach."

—— Prof Lord Broers, former Vice-Chancellor of the University of Cambridge

"The Macat vision is exceptionally exciting. It focuses upon new modes of learning which analyse and explain seminal texts which have profoundly influenced world thinking and so social and economic development. It promotes the kind of critical thinking which is essential for any society and economy. This is the learning of the future."

—— Rt Hon Charles Clarke, former UK Secretary of State for Education

"The Macat analyses provide immediate access to the critical conversation surrounding the books that have shaped their respective discipline, which will make them an invaluable resource to all of those, students and teachers, working in the field."

—— Prof William Tronzo, University of California at San Diego

The Macat Library
世界思想宝库钥匙丛书

TITLE	中文书名	类别
An Analysis of Arjun Appadurai's *Modernity at Large: Cultural Dimensions of Globalization*	解析阿尔君·阿帕杜莱《消失的现代性：全球化的文化维度》	人类学
An Analysis of Claude Lévi-Strauss's *Structural Anthropology*	解析克劳德·列维-斯特劳斯《结构人类学》	人类学
An Analysis of Marcel Mauss's *The Gift*	解析马塞尔·莫斯《礼物》	人类学
An Analysis of Jared M. Diamond's *Guns, Germs, and Steel: The Fate of Human Societies*	解析贾雷德·M.戴蒙德《枪炮、病菌与钢铁：人类社会的命运》	人类学
An Analysis of Clifford Geertz's *The Interpretation of Cultures*	解析克利福德·格尔茨《文化的解释》	人类学
An Analysis of Philippe Ariès's *Centuries of Childhood: A Social History of Family Life*	解析菲力浦·阿利埃斯《儿童的世纪：旧制度下的儿童和家庭生活》	人类学
An Analysis of W. Chan Kim & Renée Mauborgne's *Blue Ocean Strategy*	解析金伟灿/勒妮·莫博涅《蓝海战略》	商业
An Analysis of John P. Kotter's *Leading Change*	解析约翰·P.科特《领导变革》	商业
An Analysis of Michael E. Porter's *Competitive Strategy: Techniques for Analyzing Industries and Competitors*	解析迈克尔·E.波特《竞争战略：分析产业和竞争对手的技术》	商业
An Analysis of Jean Lave & Etienne Wenger's *Situated Learning: Legitimate Peripheral Participation*	解析琼·莱夫/艾蒂纳·温格《情境学习：合法的边缘性参与》	商业
An Analysis of Douglas McGregor's *The Human Side of Enterprise*	解析道格拉斯·麦格雷戈《企业的人性面》	商业
An Analysis of Milton Friedman's *Capitalism and Freedom*	解析米尔顿·弗里德曼《资本主义与自由》	商业
An Analysis of Ludwig von Mises's *The Theory of Money and Credit*	解析路德维希·冯·米塞斯《货币和信用理论》	经济学
An Analysis of Adam Smith's *The Wealth of Nations*	解析亚当·斯密《国富论》	经济学
An Analysis of Thomas Piketty's *Capital in the Twenty-First Century*	解析托马斯·皮凯蒂《21世纪资本论》	经济学
An Analysis of Nassim Nicholas Taleb's *The Black Swan: The Impact of the Highly Improbable*	解析纳西姆·尼古拉斯·塔勒布《黑天鹅：如何应对不可预知的未来》	经济学
An Analysis of Ha-Joon Chang's *Kicking Away the Ladder*	解析张夏准《富国陷阱：发达国家为何踢开梯子》	经济学
An Analysis of Thomas Robert Malthus's *An Essay on the Principle of Population*	解析托马斯·罗伯特·马尔萨斯《人口论》	经济学

An Analysis of John Maynard Keynes's *The General Theory of Employment, Interest and Money*	解析约翰·梅纳德·凯恩斯《就业、利息和货币通论》	经济学
An Analysis of Milton Friedman's *The Role of Monetary Policy*	解析米尔顿·弗里德曼《货币政策的作用》	经济学
An Analysis of Burton G. Malkiel's *A Random Walk Down Wall Street*	解析伯顿·G. 马尔基尔《漫步华尔街》	经济学
An Analysis of Friedrich A. Hayek's *The Road to Serfdom*	解析弗里德里希·A. 哈耶克《通往奴役之路》	经济学
An Analysis of Charles P. Kindleberger's *Manias, Panics, and Crashes: A History of Financial Crises*	解析查尔斯·P. 金德尔伯格《疯狂、惊恐和崩溃：金融危机史》	经济学
An Analysis of Amartya Sen's *Development as Freedom*	解析阿马蒂亚·森《以自由看待发展》	经济学
An Analysis of Rachel Carson's *Silent Spring*	解析蕾切尔·卡森《寂静的春天》	地理学
An Analysis of Charles Darwin's *On the Origin of Species: by Means of Natural Selection, or The Preservation of Favoured Races in the Struggle for Life*	解析查尔斯·达尔文《物种起源》	地理学
An Analysis of World Commission on Environment and Development's *The Brundtland Report: Our Common Future*	解析世界环境与发展委员会《布伦特兰报告：我们共同的未来》	地理学
An Analysis of James E. Lovelock's *Gaia: A New Look at Life on Earth*	解析詹姆斯·E. 拉伍洛克《盖娅：地球生命的新视野》	地理学
An Analysis of Paul Kennedy's *The Rise and Fall of the Great Powers: Economic Change and Military Conflict from 1500–2000*	解析保罗·肯尼迪《大国的兴衰：1500—2000 年的经济变革与军事冲突》	历史
An Analysis of Janet L. Abu-Lughod's *Before European Hegemony: The World System A. D. 1250–1350*	解析珍妮特·L. 阿布-卢格霍德《欧洲霸权之前：1250—1350 年的世界体系》	历史
An Analysis of Alfred W. Crosby's *The Columbian Exchange: Biological and Cultural Consequences of 1492*	解析艾尔弗雷德·W. 克罗斯比《哥伦布大交换：1492 以后的生物影响和文化冲击》	历史
An Analysis of Tony Judt's *Postwar: A History of Europe since 1945*	解析托尼·朱特《战后欧洲史》	历史
An Analysis of Richard J. Evans's *In Defence of History*	解析理查德·J. 艾文斯《捍卫历史》	历史
An Analysis of Eric Hobsbawm's *The Age of Revolution: Europe 1789–1848*	解析艾瑞克·霍布斯鲍姆《革命的年代：欧洲 1789—1848 年》	历史

An Analysis of Roland Barthes's *Mythologies*	解析罗兰·巴特《神话学》	文学与批判理论
An Analysis of Simone de Beauvoir's *The Second Sex*	解析西蒙娜·德·波伏娃《第二性》	文学与批判理论
An Analysis of Edward W. Said's *Orientalism*	解析爱德华·W. 萨义德《东方主义》	文学与批判理论
An Analysis of Virginia Woolf's *A Room of One's Own*	解析弗吉尼亚·伍尔芙《一间自己的房间》	文学与批判理论
An Analysis of Judith Butler's *Gender Trouble*	解析朱迪斯·巴特勒《性别麻烦》	文学与批判理论
An Analysis of Ferdinand de Saussure's *Course in General Linguistics*	解析费尔迪南·德·索绪尔《普通语言学教程》	文学与批判理论
An Analysis of Susan Sontag's *On Photography*	解析苏珊·桑塔格《论摄影》	文学与批判理论
An Analysis of Walter Benjamin's *The Work of Art in the Age of Mechanical Reproduction*	解析瓦尔特·本雅明《机械复制时代的艺术作品》	文学与批判理论
An Analysis of W. E. B. Du Bois's *The Souls of Black Folk*	解析 W.E.B. 杜波依斯《黑人的灵魂》	文学与批判理论
An Analysis of Plato's *The Republic*	解析柏拉图《理想国》	哲学
An Analysis of Plato's *Symposium*	解析柏拉图《会饮篇》	哲学
An Analysis of Aristotle's *Metaphysics*	解析亚里士多德《形而上学》	哲学
An Analysis of Aristotle's *Nicomachean Ethics*	解析亚里士多德《尼各马可伦理学》	哲学
An Analysis of Immanuel Kant's *Critique of Pure Reason*	解析伊曼努尔·康德《纯粹理性批判》	哲学
An Analysis of Ludwig Wittgenstein's *Philosophical Investigations*	解析路德维希·维特根斯坦《哲学研究》	哲学
An Analysis of G. W. F. Hegel's *Phenomenology of Spirit*	解析 G. W. F. 黑格尔《精神现象学》	哲学
An Analysis of Baruch Spinoza's *Ethics*	解析巴鲁赫·斯宾诺莎《伦理学》	哲学
An Analysis of Hannah Arendt's *The Human Condition*	解析汉娜·阿伦特《人的境况》	哲学
An Analysis of G. E. M. Anscombe's *Modern Moral Philosophy*	解析 G. E. M. 安斯康姆《现代道德哲学》	哲学
An Analysis of David Hume's *An Enquiry Concerning Human Understanding*	解析大卫·休谟《人类理解研究》	哲学

English Title	Chinese Title	Category
An Analysis of Søren Kierkegaard's *Fear and Trembling*	解析索伦·克尔凯郭尔《恐惧与战栗》	哲学
An Analysis of René Descartes's *Meditations on First Philosophy*	解析勒内·笛卡尔《第一哲学沉思录》	哲学
An Analysis of Friedrich Nietzsche's *On the Genealogy of Morality*	解析弗里德里希·尼采《论道德的谱系》	哲学
An Analysis of Gilbert Ryle's *The Concept of Mind*	解析吉尔伯特·赖尔《心的概念》	哲学
An Analysis of Thomas Kuhn's *The Structure of Scientific Revolutions*	解析托马斯·库恩《科学革命的结构》	哲学
An Analysis of John Stuart Mill's *Utilitarianism*	解析约翰·斯图亚特·穆勒《功利主义》	哲学
An Analysis of Aristotle's *Politics*	解析亚里士多德《政治学》	政治学
An Analysis of Niccolò Machiavelli's *The Prince*	解析尼科洛·马基雅维利《君主论》	政治学
An Analysis of Karl Marx's *Capital*	解析卡尔·马克思《资本论》	政治学
An Analysis of Benedict Anderson's *Imagined Communities*	解析本尼迪克特·安德森《想象的共同体》	政治学
An Analysis of Samuel P. Huntington's *The Clash of Civilizations and the Remaking of World Order*	解析塞缪尔·P.亨廷顿《文明的冲突与世界秩序的重建》	政治学
An Analysis of Alexis de Tocqueville's *Democracy in America*	解析阿列克西·德·托克维尔《论美国的民主》	政治学
An Analysis of John A. Hobson's *Imperialism: A Study*	解析约翰·A.霍布森《帝国主义》	政治学
An Analysis of Thomas Paine's *Common Sense*	解析托马斯·潘恩《常识》	政治学
An Analysis of John Rawls's *A Theory of Justice*	解析约翰·罗尔斯《正义论》	政治学
An Analysis of Francis Fukuyama's *The End of History and the Last Man*	解析弗朗西斯·福山《历史的终结与最后的人》	政治学
An Analysis of John Locke's *Two Treatises of Government*	解析约翰·洛克《政府论》	政治学
An Analysis of Sun Tzu's *The Art of War*	解析孙武《孙子兵法》	政治学
An Analysis of Henry Kissinger's *World Order: Reflections on the Character of Nations and the Course of History*	解析亨利·基辛格《世界秩序》	政治学
An Analysis of Jean-Jacques Rousseau's *The Social Contract*	解析让-雅克·卢梭《社会契约论》	政治学

An Analysis of Odd Arne Westad's *The Global Cold War: Third World Interventions and the Making of Our Times*	解析文安立《全球冷战：美苏对第三世界的干涉与当代世界的形成》	政治学
An Analysis of Sigmund Freud's *The Interpretation of Dreams*	解析西格蒙德·弗洛伊德《梦的解析》	心理学
An Analysis of William James' *The Principles of Psychology*	解析威廉·詹姆斯《心理学原理》	心理学
An Analysis of Philip Zimbardo's *The Lucifer Effect*	解析菲利普·津巴多《路西法效应》	心理学
An Analysis of Leon Festinger's *A Theory of Cognitive Dissonance*	解析利昂·费斯汀格《认知失调论》	心理学
An Analysis of Richard H. Thaler & Cass R. Sunstein's *Nudge: Improving Decisions about Health, Wealth, and Happiness*	解析理查德·H.泰勒/卡斯·R.桑斯坦《助推：如何做出有关健康、财富和幸福的更优决策》	心理学
An Analysis of Gordon Allport's *The Nature of Prejudice*	解析高尔登·奥尔波特《偏见的本质》	心理学
An Analysis of Steven Pinker's *The Better Angels of Our Nature: Why Violence Has Declined*	解析斯蒂芬·平克《人性中的善良天使：暴力为什么会减少》	心理学
An Analysis of Stanley Milgram's *Obedience to Authority*	解析斯坦利·米尔格拉姆《对权威的服从》	心理学
An Analysis of Betty Friedan's *The Feminine Mystique*	解析贝蒂·弗里丹《女性的奥秘》	心理学
An Analysis of David Riesman's *The Lonely Crowd: A Study of the Changing American Character*	解析大卫·理斯曼《孤独的人群：美国人社会性格演变之研究》	社会学
An Analysis of Franz Boas's *Race, Language and Culture*	解析弗朗兹·博厄斯《种族、语言与文化》	社会学
An Analysis of Pierre Bourdieu's *Outline of a Theory of Practice*	解析皮埃尔·布尔迪厄《实践理论大纲》	社会学
An Analysis of Max Weber's *The Protestant Ethic and the Spirit of Capitalism*	解析马克斯·韦伯《新教伦理与资本主义精神》	社会学
An Analysis of Jane Jacobs's *The Death and Life of Great American Cities*	解析简·雅各布斯《美国大城市的死与生》	社会学
An Analysis of C. Wright Mills's *The Sociological Imagination*	解析C.赖特·米尔斯《社会学的想象力》	社会学
An Analysis of Robert E. Lucas Jr.'s *Why Doesn't Capital Flow from Rich to Poor Countries?*	解析小罗伯特·E.卢卡斯《为何资本不从富国流向穷国？》	社会学

An Analysis of Émile Durkheim's *On Suicide*	解析埃米尔·迪尔凯姆《自杀论》	社会学
An Analysis of Eric Hoffer's *The True Believer: Thoughts on the Nature of Mass Movements*	解析埃里克·霍弗《狂热分子：群众运动圣经》	社会学
An Analysis of Jared M. Diamond's *Collapse: How Societies Choose to Fail or Survive*	解析贾雷德·M. 戴蒙德《大崩溃：社会如何选择兴亡》	社会学
An Analysis of Michel Foucault's *The History of Sexuality Vol. 1: The Will to Knowledge*	解析米歇尔·福柯《性史（第一卷）：求知意志》	社会学
An Analysis of Michel Foucault's *Discipline and Punish*	解析米歇尔·福柯《规训与惩罚》	社会学
An Analysis of Richard Dawkins's *The Selfish Gene*	解析理查德·道金斯《自私的基因》	社会学
An Analysis of Antonio Gramsci's *Prison Notebooks*	解析安东尼奥·葛兰西《狱中札记》	社会学
An Analysis of Augustine's *Confessions*	解析奥古斯丁《忏悔录》	神学
An Analysis of C. S. Lewis's *The Abolition of Man*	解析C. S. 路易斯《人之废》	神学

图书在版编目（CIP）数据

解析蕾切尔·卡森《寂静的春天》: 汉、英 / 妮姬·斯普林格（Nikki Springer）著；祝平译.—上海：上海外语教育出版社,2020
（世界思想宝库钥匙丛书）
ISBN 978-7-5446-6392-2

Ⅰ.①解… Ⅱ.①妮… ②祝… Ⅲ.①环境保护－普及读物－汉、英 Ⅳ.①X-49

中国版本图书馆CIP数据核字(2020)第056444号

This Chinese-English bilingual edition of *An Analysis of Rachel Carson's Silent Spring* is published by arrangement with Macat International Limited.
Licensed for sale throughout the world.

本书汉英双语版由Macat国际有限公司授权上海外语教育出版社有限公司出版。
供在全世界范围内发行、销售。

图字：09 - 2018 - 549

出版发行: **上海外语教育出版社**
（上海外国语大学内） 邮编：200083
电　　话: 021-65425300（总机）
电子邮箱: bookinfo@sflep.com.cn
网　　址: http://www.sflep.com
责任编辑: 梁瀚杰

印　　刷: 上海叶大印务发展有限公司
开　　本: 890×1240　1/32　印张 5.625　字数 115千字
版　　次: 2020年10月第1版　2020年10月第1次印刷
印　　数: 2 100 册

书　　号: ISBN 978-7-5446-6392-2
定　　价: 30.00元

本版图书如有印装质量问题，可向本社调换
质量服务热线: 4008-213-263　电子邮箱: editorial@sflep.com